Stargazers
The Contribution of
Amateurs to Astronomy

S. Dunlop M. Gerbaldi (Eds.)

Stargazers

The Contribution of Amateurs to Astronomy

Proceedings of Colloquium 98 of the IAU,
June 20–24, 1987

With 43 Figures

Springer-Verlag Berlin Heidelberg New York
London Paris Tokyo

Storm Dunlop

140 Stocks Lane, East Wittering, West Sussex,
PO20 8NT England

Dr. Michèle Gerbaldi

Centre National de la Recherche Scientifique,
Institut d'Astrophysique, 98 bis bd. Arago, F-75014 Paris, France

---

Photograph taken by M. de Kerolyr at the Station of Astrophysics of the Observatoire de Paris (Forclalquier - Alpes de Haute Provence)
16 January 1934 - exposure time 3 hours - plate Opta.

Collection of the Observatoire de Paris - Reproduction by F. Warin, Institut d'Astrophysique de Paris

---

ISBN-13:978-3-540-50230-2 e-ISBN-13:978-3-642-74020-6
DOI: 10.1007/978-3-642-74020-6

2156/3150-543210 - Printed on acid-free paper

Editors' Preface

IAU Colloquium No. 98 "The Contribution of Amateurs to Astronomy" was held in Paris, at the "Chaillot-Galliera" Conference Centre, on 20–24 June, 1987. A total of 250 participants – professional and amateur astronomers – came from 27 different countries. It was the first time that such a meeting had been organized under the auspices of the International Astronomical Union, and the arrangements in Paris were made by the Société Astronomique de France, which was founded one hundred years previously by Camille Flammarion, so the 100th anniversary of the SAF was also celebrated. Nine commissions of the IAU co-sponsored the meeting; it also received support from the following associations:
AAVSO (American Association of Variable Star Observers), BAA (British Astronomical Association), IOTA (International Occultation Timing Association), SRBA (Société Royale Belge d'Astronomie).

The Chairman of the Scientific Organizing Committee was Prof. J.-C. Pecker, and the names of other members who shared in the work of organizing the Colloquium are:
Dr P. Couteau (France), Cmdr H.D. Howse (U.K.), Dr M. Gerbaldi (France), Prof. J. Kleczek (Czechoslovakia), Dr I. Kozai (Japan), Dr J. Mattei (U.S.A.), Dr S. Nakano (Japan), Dr G. Taylor (U.K.), Dr V. Trimble (U.S.A.)

The Chairman of the Local Organizing Committee was Dr P. Simon, the then President of the SAF, who organized all the events in Paris, helped by P. de la Cotardière and A. Chenevez.

The proceedings were conducted in French and English and the 12 sessions were chaired by:
A.C. Levasseur-Regourd (Session 11), A. Koeckelenbergh (12), S. Debarbat (21), J. Kleczek (22), J. Mattei (23), J.-C. Merlin (24), P. Maley (25), F. Wood (26), I. Garcia de la Rosa (27), J. Rösch (31), Y. Dargery (32), A. Acker (33).

A total of 128 invited and contributed communications were given, and these *Proceedings* "espèrent refléter ces débats".

This Colloquium highlighted the fact that in several domains (most especially those of comets, variable stars, novae, etc.) the amateur contribution is becoming "irremplaçable". A second fact that became very clear is the important role played by amateurs in popularizing astronomy. In addition, this Colloquium emphasized how strong and efficient the links between professional and amateur astronomers can be.

It can justifiably be said that "Ce colloque est une grande première, couronnée de succès."

This Colloquium's *Proceedings* differ from those of its predecessors in several ways. As editors we were faced with the usual dilemma of whether to publish an incomplete volume or to delay publication by waiting for late papers and chasing them. We adopted the second method (for various reasons) and apologize to those authors who supplied their papers "at the right time". Not the least of our problems was the impossibility of publishing all 128 contributions in full, in English. As approximately half the contributions were in French, such a step would have involved a great deal of time and labour in translation, as well as resulting in a large book that would be far too expensive for most of the intended readers to afford. We have therefore ensured that most coverage has been given to those papers that detail new methods of observation, significant results of amateur work or amateur/professional cooperation, or that make suggestions for future research projects. A few papers have been withdrawn by their authors, for various reasons, such as a similar item having been previously published elsewhere. All other contributions are, however, included in at least abstract form and, where possible, appropriate references have been given. The majority of the abstracts are new, being rewritten to reflect the material that was actually presented at the Colloquium.

In compiling this record of the proceedings of the Colloquium we have not followed the chronological sequence in detail. For reasons of space, no posters are included. All the contributions are being published in full, in French, in special issues of the 'Observations et Travaux' of the Société Astronomique de France. Even where we have had to include shorter versions or abstracts in this volume, the full text submitted by many contributors has greatly helped us in editing the material or writing new abstracts. In those instances where no material was submitted by the author, the abstracts published in the original programme have been used as a basis for the reports. In a few cases where an author's mother tongue was neither French nor English, we have taken the liberty of editing the material to ensure that it is readily comprehensible. In those contributions that make reference to specific books we have, where possible, given details of English-language editions.

The translation of the contributions into English has been undertaken by one of us (S.D.), who would like to thank Mr Peter Hingley, Librarian of the Royal Astronomical Society, for his assistance in checking and locating specific references and publications in various languages.

Regrettably, it did not prove possible to use much of the material that was submitted in nominally 'camera-ready' form. We are therefore particularly grateful to Mme C. Boillet of the Institut d'Astrophysique de Paris for retyping many contributions. We should also like to record our thanks to André Chenevez, Philippe de la Cotardière, Derek McNally, Jean-Claude Pecker and Paul Simon for their helpful discussion of various matters relating to the publication of these *Proceedings*.

Finally, it remains for us to thank the sponsoring organisations and the various individuals involved with the Colloquium. First, we acknowledge the help of the IAU in the form of a grant to the SAF to support the Colloquium. The SAF itself financed the organization of the Colloquium.

We thank Prof. Maurice Lévy, Président of the Cité des Sciences et de l'Industrie for his welcome, as well as for the free planetarium show and visit to the Cité that he arranged for participants.

We thank all the astronomers of the Observatoire de Meudon, who organized the visit by Colloquium participants to the observatory, and who acted as guides to the equipment and facilities. We are also grateful to the Observatory for arranging for the closing dinner to be held there.

We similarly thank the Président of the Observatoire de Paris, Dr P. Charvin for the welcome to the observatory and the cocktail party given to Colloquium participants, and to astronomers of the Observatoire de Paris who guided the visitors round the "Carte du Ciel" exhibition organized by the observatory. Our thanks are also due to Madame Aurillac, Conseiller de Paris, for her welcome at the reception held at the Hôtel de Paris.

Together with the Société de la Tour Eiffel, the SAF organized the first international exhibition for amateur astronomers, and our thanks also go to M Arrivetz and the Société Imprimerie Spéciale de Banque, Paris for their sponsorship in the form of the printing of various items for the Colloquium.

We are grateful to all the persons who participated in the secretarial work involved with the Colloquium, mailed material to the participants, ran the registration desk, projected the occasional slide, etc.: Mmes Blomme, Bousquet, Caillat, Duval, Hedreul, Meriot, Naddeo, Pierron, and Regnault; Mlles Lericque and Marigny; MM Fournon, Louis-Tisserand, Pezzana, Valls-Gabaud, and Zimmermann. We are grateful to Mme Delmas who has helped us in many respects.

But, in fact, all these efforts were orchestrated by the Société Astronomique de France, without whom nothing could have been realized.

Michèle Gerbaldi, I.A.P., Paris
Storm Dunlop, Chichester

Introductory Address

Prof. J.-C. Pecker
Chairman of the Scientific Organizing Committee

Ladies and Gentlemen, Colleagues,

Astronomy only became a profession late in history. The existence of the profession of astronomer can be dated from the 17th century and the creation of the observatories at Greenwich, at Paris, and elsewhere. According to its statute of 1699, the Académie de Sciences (in Paris) had an astronomical section consisting of three members. We should note that it had decided, at the time of its foundation by Colbert in 1666, to debar astrology, which was still respected in the previous century, from being considered among the true sciences.

Until then, the great names in astronomy had lived on their own private incomes, or even on the taxes that they raised, like Tycho Brahe; or on the sale of horoscopes, like Kepler; or from their ecclesiastical income, like Copernicus. If they found a rich protector, it was he who financed everything and who gained the glory, even to the point of giving his name to catalogues: we may recall the 13th-century Alphonsine Tables in Spain, or Kepler's Rudolphine Tables.

But even after there were functioning observatories, their incumbents remained "amateurs", and only rarely dipped into the royal purse in order to carry out their astronomical duties.

And – in an essentially continuous fashion – running parallel to the research carried out by the professionals in their observatories, numerous amateur astronomers carried out observations and calculations at a comparable level of sophistication.

This parallel development naturally ceased with the appearance of very large instruments: how could amateurs rival Palomar or even Saint-Michel? But observing time on large instruments is expensive and is only given parsimoniously; so much so that for professionals the advent of giant telescopes meant that certain types of research had to be abandoned. Long-period variable stars? Finished. Systematic research about comets or asteroids? You don't get enough publications out of that, etc. In short, a new niche became obvious, twenty or thirty years ago, for amateur astronomers again to contribute to the progress of scientific knowledge, on the same terms as professionals. Having thus become, once again, "honorary astronomers" – a term that perhaps describes their activity better than any other – they are very numerous and found all over the world. Their relations with professionals, often even in the heart of the International Astronomical Union itself, have become more regular and more confident. It is now a question of working relationships, and many honorary astronomers are now contributing to the development of our science and work in the major observatories and university laboratories.

It therefore seemed that the time had come for a major conference, a major colloquium, to be devoted to this new symbiosis. In France, the Société Astronomique de France plays a primary role in this symbiosis – does not its Council include among the amateur astronomers a considerable number of professionals? Founded in 1887 by Camille Flammarion, it is celebrating its centenary this year. It was an ideal occasion; and the Scientific Committee would like to express its best wishes to this youthful "old lady" for a wonderful second century. The following associations joined with the SAF in acting as god-parents for this colloqium: the American Association of Variable Star Observers, the British Astronomical Association, the International Occultation Timing Association, and the Société Royale Belge d'Astronomie.

For the benefit of English-speaking participants, I shall first comment on this obviously French word, "amateur", which unites us today. I have mentioned that in French, as in English, the word has a bad connotation, but a wrong one: amateur astronomers are of course not amateurish, indeed they do *love* astronomy, as the French word also has this meaning (to love: aimer, in French). That is certainly what we all have in common, our love for the things of the sky, our enthusiasm for celestial spectacles, from eclipses to supernovae, from comets to occultations, our great happiness in understanding these marvels, our passion for the Sun, the stars, the Galaxy. Love indeed.

Many people, guided by this love for astronomy, are here today, coming from all parts of the world. I ask them to be indulgent to us. Not only because this meeting was difficult to organize; you may find flaws; I hope you will pardon us for them. Often also, we shall speak French, as much as English. It is of course easier to express oneself in one's own language. It would be better still if all of us were able to speak the twenty languages spoken in this room... English and French are the statutory working languages of the IAU – hence we have limited the linguistic scope to these two – with apologies to the others!

The International Astronomical Union has agreed to this meeting being an IAU Colloquium, number 98. I am grateful to the Executive Committees that gave such a warm welcome to our proposal, just as I thank the Presidents of ten IAU Commissions that agreed to support this project. Their suggestions resulted in the formation of the Scientific Organizing Committee, to the members of which I express my grateful thanks for their invaluable help and advice.

But I must not delay the scientific talks and discussions, which I also await with great impatience. Thank you all for coming and for your participation. I am certain that this meeting will result in many new relationships, and fertile and promising friendships.

Bienvenue et bon travail!

Welcome!

Table of Contents

Part I Historical

Amateur Astronomy from Its Origins to Camille Flammarion
J. Seidengart 2
Camille Flammarion (1842–1925): Founder of the Société Astronomique de France
J. Pernet 11
The Historical Contribution of Amateurs to the Study of Double Stars
P. Couteau 15
Twentieth-Century Amateur Astronomy
A.J. Meadows 20
Criteria for Identifying an Astronomer as an Amateur
T.R. Williams 24
Johannes Hevelius: Polish Seventeenth-Century Brewer and Astronomer
C. Iwaniszewska 26
Eugène Antoniadi (1870–1944) and Planetary Observation
A. Dollfus 28
Investigating Astronomy's History
N. Sperling 31
Charles Boyer and the Rotation of Venus
A. Dollfus 32
The Amateur and Eclipsing Binary Stars
F.B. Wood 35
Amateur Astronomy in Poland: Past and Present
K. Turaj 40
Giovanni Battista Lacchini: An Amateur Astronomer from Italy
G. Favero, S. Baroni 44
The Instruments of H.E. Dall
E.J. Hysom 46
Roberts of Lovedale and Eclipsing Binary Stars
T.R. Williams 48

Seventeenth-Century Solar Observations: Fundamental Results with Amateur Instruments
E. Ribes-Nesme, J.C. Ribes 50
Sailors and Astronomy
A. Demerliac 51
The Map of the Heavens at the Chateau de Saint-Jean-de-Chepy
J. Barbier 54
Amateur Astronomy in France, 1789–1830: Two Examples
S. Dumont, S. Grillet 55
Some French Amateurs of the 2nd Half of the 19th Century
F. Hourmat 56
Comet Halley in 1910 in Franche-Comté: or Collective Memory at Fault
A. Heurtel, M. Rougeot, J.P. Parisot 57
A Renowned, Yet Forgotten Astronomer: Marcel de Kerolyr
A. Richardot 58
Historical Session: The Contribution of Amateurs to Astronomy, Yesterday and Today
A. Koeckelenbergh 62

Part II Observational Methods

Amateur Astronomers and the IAU Central Bureau for Astronomical Telegrams and Minor Planet Center
B.G. Marsden 64
Contributions of Amateur Astronomers to Variable-Star Observing
J.A. Mattei 77
Observations of Comets, Minor Planets and Meteors in Japan
I. Hasegawa 87
Amateur Astronomers' Contribution to the HIPPARCHOS Programme
P. Bacchus 90
The S.A.F. Double Star Section's Contribution to the INCA-HIPPARCHOS Programme
P. Durand 93
Working with Amateur Astronomers
S.J. Edberg 95
Amateur Astronomers and the Hubble Space Telescope
S.J. Edberg 100
The T.60 Operation at Pic du Midi
C. Buil, O. Las Vergnas 102
The Effects of Light Pollution on Amateur Astronomy
D. Crawford 104
An Amateur CCD Camera
C. Buil, G. Prat, E. Thouvenot 105

A Spectrograph for the T.60 at Pic du Midi
D. Bardin, D. Buty 108
Using an Image-Intensifier in Astronomy
T. Midavaine 110
Amateur Polarimetric Research
D. Böhme 110
Searching for Variables, and Constructing a Slit-less Spectroscope
L. Dahlmark 111
Solar Spectroscopy with a CCD
O. Saint-Pé 111
Obtaining Extraterrestrial Dust by Balloon Probes
L. Baufette 111
Making a Telescope Mirror
J.M. Becker 112
Searching for Comets
B. Mayer 112
Achievements in Astronomical Video Data Collection
P.D. Maley 113
Video Grazing Occultations
D.W. Dunham 120
An Improved Method of Photographic Measures of Double Stars
D. Rodriguez, J.A. Bernedo, D. Valls-Gabaud 122
The Scientific Merit of Amateur Astrophotography
W.E. Celnik, P. Riepe 128
Reduction of Variable-Star Observations Using Basicode
T. Jurriens 131
Image Digitizer
P.A. Magnien 133
"Comet" Software
P. Butzbach, J.P. Parisot 133
Refining the Orbit of Visual Double Stars
E. Soulie 133
Reliability and Accuracy of Astronomical Observations by Amateurs
Z. Kviz 134
Trying to Reduce Errors in Visual Estimates of Variable Star Magnitudes
E. Velasco, P. Velasco 138

Part III Observations and Results

Photographic Observation of Stellar Occultations
A. Tomić, N. Čabrić, V. Čelebonović 144
Studying "Luna Incognita": The Region Near the South Pole
M. Legrand 146

Asteroid and Cometary Occultations
D.W. Dunham 147
Astrometric Measurements of Rarely-Observed Minor Planets
R. Chanal 149
Video Recordings of Asteroidal Occultations
R. Boninsegna, J. Lecacheux 149
Space Exploration of Comets and the Evolution of Amateur-Professional Cooperation
A.-C. Levasseur-Regourd 150
The Contribution of Amateur Astronomers to Cometary Observation
A. Heck 152
New Method of Statistical Analysis of Cometary Light-Curves Applied to P/Halley
S. Thebault 153
P/Halley: The Disconnection Event of 1986 April 11
P.M. Bergé, G. Mahoux, A.C. Levasseur-Regourd 155
A Study of Disconnection Velocities in the Plasma Tail of P/Halley
R. Caron, S. Koutchmy, M. Sarrazin 158
Analysis of the Light-Curve of Comet P/Halley in 1985–1986
D. Valls-Gabaud, A. Gómez 160
New Trends in the Discovery of Comets by Amateurs
J.C. Merlin 163
Amateur Astronomers and the Recovery of Periodic Comets
P. Martinez 163
Shape and Frequency of Outbursts of P/Schwassmann-Wachmann 1
J.C. Merlin, D. Fayard, E. Gérard 164
Astrometry of Comet Halley
H. Hadavi 164
Observation and Analysis of Comet Tails
R.L.W. van de Weg 164
Photometric Analysis of the Solar Corona on 1984 November 22/23
R. Caron, Ch. Nitscheim, M. Sarrazin 166
Chromospheric Eruptions Observed with a Wide Slit
F. Costard 168
Amateur/Professional Cooperation in 2 Solar Studies
T. Roudier, R. Muller, J.C. Hulot, F. Vaissière 168
The IOTA Expedition Results from the Total-Annular Eclipse in Gabon, 1987
P.D. Maley 169
Meteor Studies
B.A. Lindblad 170
The “Sonate” Campaign
J. Mazel, R. Futaully 173
Two Solar Observation Programmes
M.J. Marthes, J.M. Malherbe, J. Vial 175

Long-term Monitoring of Solar Activity by Amateur Astronomers
K. Reinsch ... 177
The Moonwatch Program: A Model for Amateur Contributions to the ISY
J. Cornell ... 181
Study of Variable Stars
E. Poretti ... 183
The Contribution of Amateur Astronomers to the Study of Variable Stars
D. Proust, E. Schweitzer ... 185
Professional/Amateur Cooperation in the Study of Variable Stars as Illustrated by the Fourth Edition of the GCVS
M.S. Frolov ... 188
Amateurs and the Search for Supenovae
S.H. Lucas ... 190
The U.K. Nova/Supernova Search Programme
G.M. Hurst ... 192
GEOS – Groupe Europeen d'Observations Stellaires
M. Dumont ... 194
GEOS Results for PX Cephei
R. Boninsegna ... 196
The French Network for Observing Comets and Small Solar-System Bodies
R. Chemin ... 197
Amateurs' Contribution to the PHEMU85 Campaign Observing Mutual Phenomena of Jupiter's Satellites
J.E. Arlot ... 198
Wanted: An International Jovian-Satellite-Phenomenon Centre
N.P. Wieth-Knudsen ... 199
The European Section of IOTA: Scientific Programme Coordination and Results
W. Breisker, H.J. Bode ... 199
The Role of the Amateur in Modern Astronomy
H.D. Howse, P. Moore ... 200

Part IV Popularization

Meeting the Public: Popular Observatories
J. Kleczek ... 202
Comet Halley: A Media Event
J.-L. Heudier ... 205
A Planetarium for Astronomy
A. Acker ... 208
Astronomical Publications
P. de la Cotardière ... 210
Problems in Publishing a Popular Journal in a Small Country
P. Gugnon, J. Sauval ... 213

Methods of Popularizing Astronomy in Various Countries
C. Iwaniszewska 214
Opening the Sky to Everyone
J. Chanoir 215
Interesting Business in Astronomy
M.L. du Moutier 215
Preparing Astronomical Videotapes
F. Suagher et al. 216
Transmission of Astronomical Data and Information by the A.F.A. Videotext System
B. Pellequer 216
The Solar System Promenade
A. Tarnutzer 216
The Open Sky Underground
I. Beauvois, O. Las Vergnas 217
Teaching Astronomy in Schools
C. Iwaniszewska 217
Teaching Astronomy and the Vital Role of the True Amateur
H.R. Mills 217
Astronomical Education in Poland
K. Turaj 218
Astronomy and Computing in Spanish (Aragon) High Schools
A. Elipe, J. Ezpeleta, V. Lanchares 218
The Planetarium at the Palais de la Découverte
G. Oudenot 218
The Starlab Mobile Planetarium in the Mid-Pyrenees
J.-P. Chrétien, J.-P. Brunet 219
The Planetarium at a School in Nice
J. Chappelet 219
The Planetarium as a Medium for Teaching Astronomy
K. Barve 219
Twenty Years of the Japan Amateur Astronomers' Convention
S. Kimura 220
Astronomy in the Big Apple
J. Pazmino 220
The Activities of Astronomical Clubs
C. Bourdeille 220
The Flammarion Astronomical Society of Montpellier
J.-M. Faidit 221
A Decade of Organized Work in Venezuela
D. Sanchez 221
The Bochum Observatory
J.V. Feitzinger et al. 221
The Triel Observatory
J.-P. Trachier 222

The Bruno-H.-Bürgel Observatory
H. Busch 222
The Strasbourg Observatory
R. Hellot 222
Feeding the Five Thousand
M. Enock 223
Efforts by the Instituto de Astrophisica de Canarias to Publicise Astronomy
I. Garcia de la Rosa et al. 223
Tailor-Made Messages: Astronomical Information for Diverse Audiences
J. Cornell 223
The Role of the Observatoire de Paris, Meudon in Providing Astronomical Information
M. Floquet, B. Schmieder 224
Observations – Concluding Remarks
B.G. Marsden 225
International Astronomical Union – Colloquium 98 General Session
A. Koeckelenbergh 228

List of Participants 231

Part I

Historical

Amateur Astronomy from Its Origins to Camille Flammarion

Jean Seidengart
Université de Paris X, 2 rue de Rouen, F-92001 Nanterre Cedex, France

Introduction

It is not possible to sketch the history of astronomy by amateurs without first defining what the term "amateur astronomer" truly means. But we must avoid using "the benefit of hind-sight" to interpret the past in terms of a concept, the relevance of which has been perceived only during the course of the last one hundred years. In order to avoid the sophistry arising from such a purely imaginary history that is based on false conceptions, it is useful to examine amateur astronomy's precise status, since its emergence in the dawn of classical science, by outlining the type of relationship that it has had with "professional" astronomy. In doing so, we can evoke some of the major developments in amateur astronomy, the scientific contributions of which, although often of high quality, have at times reached the very forefront of the discipline of astronomy.

In early classical French, the term "amateur" referred to anyone who had a keen interest in the arts, being based primarily on the strong love that could arise for any object capable of stirring the emotion. In the course of the 17th century, the term came to be extended to cover the sciences and other fields in general. An amateur is therefore, above all, anyone who does not live by astronomy – disregarding how meagre the income of official astronomers may have been in the past. From this we may see that it has to be one thing or the other: either amateurs gain their income from some other professional activity – but they are then unable to devote their whole time to astronomy – or else they are sufficiently rich to be able to give their undivided attention to their love. In the latter case, however, they risk being just a "dilettante" (that is to say no more than a "curiosity-seeker", who is only drawn by what is unusual or new); a dilettante, moreover, whose practice of astronomy is influenced more by inclination than by the exigencies of a plan of research that has been methodically and rationally devised within an overall cooperative programme. Dilettantes are not forced to preserve any continuity in their observational programmes, the content of which may vary according to the whim of the moment.

Amateurs, therefore, have less time at their disposal than professionals, or else work in a very disjointed fashion. In most cases their initial theoretical grounding is much shallower and less technical than that of professionals, even if these same amateurs have often been professional scientists in other disciplines (mathematics, mechanics, optics, engineering science, etc.). With some rare exceptions (some of which have been of the greatest importance as we shall see in considering Herschel and Lord Rosse), amateurs have only modest technical and material resources at their disposal, and these are also frequently of poorer performance than those of professionals, especially from the 19th century onwards. Nevertheless, if we leave aside the casual "curiosity-seekers" and the "dilettantes", we are still left with the "experts", who have been capable of making skilled observations that are not only regular, but also are precise in their measurements. As we shall see, the development of astronomy owes much to them and this because, in the words of Alexander Koyré, they have been able to rise from "the world of the approximate to the Universe of the precise". Advanced amateurs, in the true sense of the term, are therefore those who have been able to go beyond the charms of simply contemplating the splendour of the heavens – which, nonetheless, remains irreplaceable – and to venture farther along the path of knowledge. Eudoxos' teacher, Plato, writing in his *Timaeus*, warned the merely curious that:

> "They came from harmless but light-witted men, who studied the heavens but imagined in their simplicity that the surest evidence in these matters comes through the eyes." (1)

It is not possible, of course, within the limits of this paper, to review exhaustively the work of amateurs between the Renaissance and modern times. To the best of our knowledge this research has never been fully undertaken, and indeed could not be carried out in any truly significant manner, because, until the end of the 18th century, the strict distinction between very advanced amateurs and officially appointed professionals hardly had any historical or epistemological relevance. Indeed, from the end of the Middle Ages until the creation of the various academies and learned societies during the classical age, it was just as likely for an "official astronomer" – the term "mathematicus" was often used – to have been a charlatan, preoccupied with astrology, or even someone more concerned with intrigue and avid for power or political influence, as for them to be a professional or a well-informed amateur. It is true that in Europe astrology was often part of the official functions of a court astronomer until the middle of the 17th century: Kepler was one of the last astronomer-astrologers of the court at Prague.

In this respect, it must be recognized that the foundation throughout Europe, during the 17th and 18th centuries, of academies and societies that could carry out collective research, did at least improve the situation to some extent. By giving an institutional status to scientific disciplines it became possible to organize and coordinate research, and to supervise the quality and level of publications. At the end of the 18th century, astronomy's autonomy was finally recognized and, as D'Alembert notes fifty years later in the *Encyclopédie*:

> "Today there are only astronomers, and no astrologers, or at least astrologers are held in very low esteem." (2)

In order to avoid any confusion over classification, we shall only begin our discussion of amateur astronomy with the scientific revolution of the 16th and 17th centuries, even though astronomy is a very much older science, and was practiced by many renowned scientists in antiquity from the time of Eudoxos to that of Ptolemy. Our treatment is determined by the amateur/professional dichotomy, the pertinence of which only appeared with the *institutionalization of science* in accordance with internal criteria of scientific competence. We shall see in the historical outline that follows that these general considerations of amateurs' contributions to the development of astronomy fall into four distinct, yet often interdependent, fields: observation, instrumentation, theory and, on the educational side, the global dissemination of astronomical knowledge.

The Dawn of the Scientific Age

With the dawn of the scientific age, the very small number of professional astronomers, whose status was still not properly defined, did have the effect of creating a considerable extension to the concept of amateur astronomy. The first "expert amateurs" at that time were specialists in other scientific disciplines (mathematicians, physicists, engineers, navigators, geographers, etc. ...), philosophers preoccupied with the question of knowledge and natural philosophy, teachers at the religious and royal colleges who often taught several disciplines at once, and self-taught men drawn from various social and professional backgrounds (artisans, business-men, military officers, clerics, aristocrats, land-owners, etc. ...). Let us note in passing that the passion for astronomy encouraged international exchange of ideas and overcame the social barriers found under the Ancien Regime, with a few isolated exceptions.

While Peiresc and Gaultier only just failed to observe the transit of Mercury across the disk of the Sun that had been predicted by Kepler, Gassendi was the only person to observe it, which he did and with precision, on 1631 November 7 from Paris. Moreover, he gave a full description in his paper of 1632, *Mercurius in sole visus*. Kepler's prediction thus had brilliant, and important, observational confirmation.

After the death of Peiresc in 1637, Gassendi continued to make observations, which can be found in volume IV of the latter's *Oeuvres Complètes*, but there is no really notable discovery amongst them. He did have the distinction of refuting the error made by Father Rheita, who thought he had discovered five new satellites of Jupiter. One remarkable result of the cooperation between these two great amateurs that is still remembered, however, is the very first map of the Moon, which they had engraved by Mellan in 1636. D'Alembert praises it a century later in the *Encyclopédie*:

> "Of all the maps of the Moon that have been published up to the present, those that were engraved by Mellan, under the supervision of Peiresc from the observations of Gassendi, are without doubt the best, and the most representative." (3)

During the same period, Father Niccolo Zucchi, S.J. (1586–1670), built, in 1616, one of the very first telescopes making use of an ocular to observe the image produced at the focus of a concave metal mirror. With this apparatus, Niccolo Zucchi observed, in 1640, the markings discovered on Mars by Fontana in 1636, and he went on to discover the existence of belts on Jupiter.

Another notable person is Fontenelle (1667–1757). This time, we are dealing with a writer devoted to the dissemination of the ideas and discoveries of the "Progressives". In this respect, the *Entretiens sur la pluralité des Mondes* (1686) represents a deliberate attempt to render the achievements of "Copernican" astronomy accessible to everyone. Even if Fontenelle had recourse to literary devices to present the material in a light-hearted manner, he still remained faithful to Cartesian epistemology. Fontenelle shows his reader that the drama of the universe should inspire us to enquire into the fundamental *mechanisms* that produce the successive scenes on stage. It is all as if the new, Copernican astronomy of Kepler and Galileo had taken us behind the scenes, and this knowledge increases the value of the very drama itself:

> "There is nothing that should interest us more", says Fontenelle, "than to know how this world that we inhabit is made, whether there are other similar worlds, and if they also are inhabited." (4)

The book was aimed at the widest possible audience, because, according to its author, it was capable of pleasing both the reader with a knowledge of physics, and the beginner:

> "I must advise those who will read this book, and who have some knowledge of physics, that I have not presumed to try to give them instruction, but rather to divert them in presenting, in a perhaps more pleasant and lighter fashion, that which they already know. To those for whom this material is new, I would say that I have tried to instruct and to amuse them at the same time. The former will be going against my intentions if they look for instruction in this book, and the latter if they only look for amusement." (5)

As far as science was concerned, Fontenelle was only an amateur, albeit well-informed in mathematics by his friend the great geometer Sauveur, in astronomy by La Hire, and in physics by Christiaan Huygens. The various successive editions of the *Entretiens* revised by Fontenelle incorporate all the new discoveries made after 1686, up to the edition of 1742, which is by far the most complete, because it adds a "sixth day", which has as a sub-title: "New thoughts that confirm the preceding discussions. The latest discoveries that have been made in the heavens." (6) The work spread across Europe like a trail of gunpowder, and brought fame to its author, who thus began a new literary genre: that of scientific popularization.

In his *Kosmotheoros* of 1698 – as famous throughout Europe at the time as Fontenelle's *Entretiens* – Huygens described the astronomical knowledge of that period, as well as his own discoveries. Huygens represents another type of amateur, quite different from Fontenelle. He was a great Dutch scientist, who worked for nearly twenty years at the Académie Royale des Sciences in Paris. He made discoveries in most of the branches of the physical and mathematical sciences, but he did not pretend to be a professional astronomer of the type represented by Hevelius, La Hire or Jean-Dominique Cassini. In astronomy for example, with his brother Constantijn he constructed refractors of very great focal length (3.5 m and 7 m) in order to avoid spherical and chromatic aberration; these had magnifications of × 100 times. Using one of his telescopes he discovered Titan, the largest satellite of Saturn, in March 1655. Then he solved the problem about the shape of Saturn, raised by Galileo in his *Siderius Nuncius*. Huygens discovered the ring surrounding the planet during the winter of 1655–1656, at a time when the "ansae" were invisible. He did not publish his discoveries until 1659, in his *Systema Saturnium*. On the instrumental side, in order to increase the field of his powerful telescopes, Huygens had the idea of placing a field lens in front of the eye lens, which on the one hand created a new type of ocular – still used nowadays and well-known to amateurs – and on the other eliminated chromatic aberration. This made it possible to increase the focal length of his refractors (going from 7 m to 15, and then to 30 and 60 m!) and thus their magnification, whilst still preserving a useful field, providing tubes, with their incurable flexure, were abandoned.

In his *Kosmotheoros*, published posthumously, Huygens described a finding that was very remarkable for its time. Suggesting that all the visible stars had approximately the same dimensions, he argued that the differences in their visual magnitudes were a function of their distance from the terrestrial observer. As Sirius had the greatest visual magnitude, he attributed its brightness, not to its actual diameter, but to the fact that it was very close to the Solar System. Pointing his telescope at the Sun, after having stopped it down by capping it with a sheet of metal through which a small hole allowed the Sun's light to pass, he blocked the hole in its turn with a bead of glass, thus reproducing the appearance of Sirius. Huygens argued that Sirius actually had the same diameter as the Sun, but that their difference in brightness was because of the extreme distance of the principal star in Canis Major. He concluded:

> "Thus, making the calculations according to the laws of dioptrics, I found that the diameter of the Sun had become 1/152 part of the fraction of 1/182 that I saw the first time through the small hole. The product of 1/152 and 1/182 is 1/27664. As a result, if we assume that Sirius is equal to the Sun, it follows that the ratio of its distance to that which separates us from the Sun is 26 664 to 1." (7)

Huygens thus anticipated the photometric method by two centuries, and even though the postulate that all stars have the same diameter was false, he nevertheless had the distinction of giving stellar distances that were an order of magnitude better than those put forward by Kepler, who spoke of 600 000 Earth diameters. Huygens' figure, although less than the actual one, was 538 times better than Kepler's. His value was given by Fontenelle in his *Entretiens* from 1708 onwards, and it can still be found up to the beginning of the 19th century!

The Eighteenth and Nineteenth Centuries

In the 18th century, Newtonian theories reached their height, and were finally accepted throughout Europe, even by the most reluctant spirits. Amateur astronomy had spread considerably among the various classes of western society (to which we restrict ourselves). Numerous well-informed amateurs were members of regional scientific societies, and even of the leading national institutions. For example, as Lesley Murdin has shown so clearly in her recent work on astronomical practice in the classical age (8), there were so many amateurs in England at the end of the 17th century that they were the subject of famous satires and lampoons at the time. Whether these amateur astronomers were of private means like Molyneux, or clergymen like William Derham and James Pound, there were more than fifty of them who sent regular communications and observations to John Flamsteed, the famous Astronomer Royal at Greenwich. In France at the beginning of the 18th century, observatories, and on a more modest scale, private observing sites, were increasing. Bigourdan, in this work (9), lists a score in the region around Paris, half of which belonged to individuals. In the provinces, the observatories belonged to educational institutions, local academies, or to individuals. There were purely private observatories or serious observing sites at Toulouse, Avignon, Marseille, Lyon, Montpellier, Mirepoix, Béziers, Saint-Lô, Viviers, Bordeaux, Montauban, Bourg-en-Bresse, etc. In astronomy it is indeed very useful to have a number of observers situated in different geographical locations well separated from one another, but it is nevertheless necessary for their observations to be coordinated and integrated within specific, methodically organized, research programmes. It was just this organization of research that was missing at that time in astronomy. In France it had to await the creation of the *Bureau des Longitudes* in 1795, for astronomical research to be organized and centralized, not to speak of the quite indispensable metrological reforms that were carried out by the *Commission des poids et mesures* under the Constituent Assembly.

Amongst all the amateur astronomers of the 18th century, one great figure stands out in the person of William Herschel (1738–1822), whose expertise and rigorous methods put him on a par with professionals. Originally a musician (he was an oboeist in the Hanoverian royal army), William Herschel emigrated to England where he developed a passionate interest in astronomy. The poor quality of telescopes that he had bought second-hand encouraged him to grind and polish mirrors himself. In 1782 he managed to complete mirrors that were far superior to those used by the official astronomers. The latter soon recognized his exceptional gift for manufacturing optical instruments and officially congratulated him about it. With one of his favourite instruments, Herschel discovered in 1781 a trans-Saturnian planet (which he took initially to be a comet), and which his friend the German astronomer J.E. Bode later called Uranus. This launched him on his true career, and he undertook the preparation of a catalogue of double stars that ended by including 848 pairs. He tried to construct a telescope 3 feet in diameter (91.44 cm) and more than 9 m long, but the mirror fractured on cooling. In 1789 he constructed, in what was an unimaginable feat at that time, a giant telescope, 1.22 m in diameter and more than 12 m long! Handling this telescope was rather difficult, but as soon as he put it into service he discovered the sixth satellite of Saturn. However, the Solar System did not interest him as much as the structure of the universe, to the study of which he devoted the rest of his life. In this respect, Herschel is the founder of modern observational cosmology.

As far as theoretical cosmology was concerned, the 18th century saw an amateur of genius, originally a philosopher, in Immanuel Kant. In 1755 the young Kant published his *Universal Natural History and Theory of the Heavens*, where the first correct conception of the structure of our Galaxy appeared, as well as the outline of what was later to be called the theory of "island-universes" (10). But Kant's hypothesis, correct though it was, remained entirely speculative and without a sufficient observational basis. It was this gap that Herschel undertook to fill, with the help of his giant telescope, the most powerful one in the world at that time. The famous astronomer Messier had, of course, already published a catalogue in 1781 that included about one hundred celestial objects that were nebular in appearance. For his part, Herschel, after twenty years of painstaking observation, described more than 2500 "nebulae" in his new catalogue of 1802. The decisive question was to determine the nature of these "nebulae". Were they clouds of dust and gas illuminated by nearby stars, or were they luminous in their own right? In other words were they not rather quite extensive clusters of stars? In the latter case, were these clusters of stars part of our Galaxy, or were they "extragalactic" in nature?

During the 1780s, Herschel put forward the view in numerous, important communications to the Royal Society, that all the "nebulae" must in fact be stellar clusters capable of being resolved into stars, with the exception of those that were too distant for the observational means available at that time. Herschel was to revise, and even totally change his opinion about the nebulae, when in November 1790 he discovered a nebula that consisted of a central star surrounded by a cloud of gas, and which is traditionally (but incorrectly) called a "planetary nebula" (11). He ended up by recognizing that this nebula is nearby and that its nebulosity could not be resolved into stars. It was equally the cause of his revising his cosmological ideas, which had a considerable influence on Laplace. Only in 1811 did W. Herschel describe to the Royal Society his evolutionary theories on cosmological subjects, in a paper of primary importance (12). He saw the nebulae as being a primitive stage in the formation of stars, which arose from a condensation of the nebular material under the force of gravity alone. The astronomer P.S. de Laplace, when he heard of Herschel's paper, saw in it obvious observational confirmation of his own cosmological theories, developed since 1796 in the various successive editions of his *Exposition du système du monde*. Laplace states in the 1813 edition of the *Exposition*:

> "The nebulae classified according to this philosophical scheme of Herschel's indicate that there is a considerable likelihood of their being transformed, at some future date, into stars. (...) Such a remarkable agreement, arrived at by following completely different routes, means that the existence of this earlier state of the Sun is very probably the truth." (13)

During the twenty years that followed the publication of his research about the nebulae, W. Herschel, who had been aided by his sister Caroline, was showered with honours by the king, George III, who appointed him astronomer to the royal family and gave him an annual pension of 200 pounds. Herschel is therefore one of the group of amateurs who became professionals.

To conclude this outline, which is really too short and too selective for such a subject, we should not fail to mention the remarkable case of William Parsons, the third Lord Rosse (1800–1867). With him we have reached the middle of the 19th century, at a time when the professional astronomers mainly used *refractors*, the accuracy and reasonable focal lengths of which – by comparison with Herschel's "monsters" – gave better service by reason of their efficiency and ease of handling. After sound scientific studies at Oxford, the future Lord Rosse undertook the construction of a *giant reflector*, despite all the difficulties presented in perfecting a bronze mirror. He intended to exceed the dimensions of Herschel's reflector, which had a diameter of 4 feet (1.22 m), with a 6-foot (1.83 m) reflector. Helped by a considerable fortune, he turned the family mansion at Birr Castle into a foundry and cast a 4-tonne mirror which took 16 weeks to cool. After some set-backs, he succeeded with his fifth mirror in April 1842. The focal length of the mirror was more than 16 metres, and after a difficult polishing process, he finally attained a resolving power of 1/10th of a second of arc, with a magnification of $\times$ 828. Lord Rosse gave his "Leviathan" a filar micrometer with phosphorescent threads in order to be able to obtain precise measurements, whilst reaching magnitude 17! The gigantic monster outclassed every other instrument in the world, and became operational in 1845. Lord Rosse and his friend, the Reverend Thomas Romney Robinson observed from 1848 to 1878, and carried out the first astronomical photography and spectroscopic studies. The results, although spectacular, did not bring all the advances that one might have expected. Lord Rosse did, however, succeed in determining the structure of the spiral arms of galaxies, notably in M 51, and also in resolving a large number of both globular and open clusters. But he was unable to resolve stars in galaxies and thus put a decisive end to the controversies that raged at the time about the nature of the nebulae. A definitive reply to this question did not come until 1924, when it was thanks to the remarkable observational work by E. Hubble (who was himself originally an amateur astronomer). Lord Rosse had not realised that it was not enough just to construct a giant telescope for it to give immediate results concomitant with its theoretical capacity, it needed to have an *adequate site*, with clear skies, which was certainly not possible on his small Irish estate. If for this reason Lord Rosse is not an incomparable observer, he is at least a highly ingenious inventor.

Conclusion

These various historical examples show us that the contribution of amateurs in the past is not really distinct from the rest of astronomy: indeed on several occasions it has been at the very forefront of research, in that heroic age when there were such pioneers. Then, in the second half of the 19th century, French amateur astronomers had the chance to join a society founded by Camille Flammarion: it was then that amateur astronomy in France achieved full status and entered a new phase, with the results that we see around us now.

References

1. Plato, in *Timaeus*, 91 D-E, trans. Cornford, F.M., London, 1937; see also *The Republic*, Book VII, 529, A-C
2. D'Alembert, *Encyclopédie*, 1756, Tome 1, p.783
3. Op. cit., cf note 2
4. Fontenelle, *Entretien sur la pluralité des Mondes*, 1742 edn, preface
5. Fontenelle, ibid.
6. Fontenelle, op. cit., ed. Marabout, after edn of 1742, p.121
7. Christiaan Huygens, *Kosmotheoros, sive De Terris Coelestibus, earumque ornatu, Conjecturae*, den Haag, 1698. [2nd English-language edition: *The Celestial Worlds Discovered...*, London, 1722 – Eds.]
8. Lesley Murdin, *Under Newton's Shadow*, Bristol & Boston, Adam Hilger, 1985
9. Bigourdan, *Histoire de l'astronomie d'observation et des observatoires en France*, Paris, 1930
10. Kant, *Allgemeine Naturgeschichte und Theorie des Himmels*, Königsberg & Leipzig, 1755. [*Universal Natural History and Theory of the Heavens*, trans. Hasties, W., Univ. Michigan Press, Ann Arbor, 1969, or the same author's *Kant's Cosmogony*, London, 1900, – Eds.]
11. This "planetary nebula" is probably NGC 1514, magnitude 10.8, situated at a distance of 4300 light-years; $\alpha 4^h 06.2^m$, δ +30° 38′.
12. c.f. *Philosophical Transactions*, 1811, p.269 et seq.
13. Laplace, *Exposition du système du monde*, 4th edition, Paris, 1813, p.431.

Camille Flammarion (1842-1925): Founder of the Société Astronomique de France

Jacques Pernet
(Curator, C. Flammarion Collection at Juvisy Observatory), 5 rue Victor Basch, F-91230 Montgeron, France

The long and fertile scientific career of Camille Flammarion, the founder of the Société Astronomique de France, whose centenary we are celebrating in 1987, is marked by an extraordinary variety of projects and research, covering practically all of the astronomical questions and problems that were current in his time. We may cite, as some of the most remarkable: the nine volumes of his *Etudes et lectures sur l'astronomie* (1867 to 1880); his *Catalogue des étoiles doubles et multiples en mouvement relatif certain* (1878), summarizing 28 000 observations; *La planète Mars* (2 vols 1892 & 1909), a priceless documentation of the first martian observations around the beginning of the 20th century; and *La planète Vénus*, a general discussion of observations (1899).

But to stay within the bounds that we have been set, and which forbid us from expanding our arguments to include an exhaustive study of Flammarion's scientific work, we shall restrict ourselves today to illustrations of his unequalled role in the birth and development of amateur astronomy, first as a fervent amateur himself and then, and primarily, as the enthusiastic innovator, who was the inspiration for innumerable people to take up astronomy as a vocation.

It is a self-evident truth to speak of Flammarion's passion for the observation of the sky. In his *Mémoires d'un astronome* (1911), he is delighted to recall that, at the earliest age, he was most impressed by two astronomical events, "as rare as they were impressive" as he describes them: two eclipses of the Sun (those of 1847 October 9 and 1851 July 28) that his mother showed him from his birth-place at Montigny-le-Roi (Haute-Marne).

So we shall not be surprised that in 1866, in Paris, when his financial situation was far from being brilliant, he did not hesitate in acquiring a 108-mm refractor, an instrument that was very respectable in size for that time. "Passionately," he says, "I started to observe and draw sunspots, the strange lunar features, the appearance of Jupiter, of Saturn, of Mars, star clusters, and double stars". And he adds a little farther on: "Those who have not tasted the pleasure, the happiness, of astronomical observation, just cannot imagine their utterly fascinating interest."

In 1882, he received, from an admirer of his work, the gift of an enormous property sited at Juvisy-sur-Orge, about twenty kilometres from Paris. His generous donor stated that he could do just as he liked with it, and even sell it again if he so wished. Flammarion did no such thing. Instead, without losing any time, and devoting the larger part of his income to it, he started to turn it into a well-equipped scientific establishment. In particular it had an excellent 240-mm equatorial, with which he, and other very talented observers, such as Antoniadi and Quénisset, obtained magnificent results.

It is obvious that Flammarion had the "sacred fire" that inspires selfless investigators. We must, moreover, not forget that from the 1880s, the popularization of astronomy, and bringing it to be understood and loved by his readers (through such introductory works as *le Merveilles célestes* 1866, *Histoire du ciel* 1872, and *Les terres du ciel* 1877), only ever appeared as the first step. In 1879, in the last pages of his famous *Astronomie populaire* he wrote: "It is not necessary to possess complicated and expensive instruments in order to begin the practical study of the sky, and we may even note that a large number of discoveries in physical astronomy have been made by ordinary amateurs, using very modest instruments." He then goes on to describe the possibilities for 61-, 75-, 95- and 108-mm refractors, respectively, and for a 100-mm reflector, adding the names and addresses of manufacturers and the price of each instrument. And he ends: "It will be seen that in our days the practice of the most beautiful of all sciences is accessible to anyone: nowadays we may well say that we have an embarrassing range of choice."

This encouragement towards a more "active" astronomy was re-emphasized in the supplement to *Astronomie populaire*, *les Etoiles et les curiosités du ciel*, which appeared the following year. After the first section devoted to the detailed description of each constellation, the second part, entitled "Various documents, instructions, tables and catalogues" essentially describes the sky for each day in the year – observations that can be made of the planets, the Moon, and the Sun; observational instruments and the practical study of the sky; a general catalogue of stars; lists of double stars and variable stars – a complete set of practical information that, collected here for beginning observers, gave them a head start.

Once he had begun, and as if to confirm the new direction of his popularization, in 1882 Flammarion started a monthly publication, *L'Astronomie*, the centenary of which we celebrated in 1982. This was a specifically astronomical magazine, and the whole project might seem somewhat premature, but it was an immediate success and the circulation was over 6000 from 1883 onwards. In the notice to his readers that appeared in the first issue, Flammarion was careful to state "Observers of the heavens – who are beginning to become quite numerous – who have made useful observations may publish them here. In each issue a specific section will be reserved for correspondence."

For the first time, amateurs had the means of getting to know one another, and to publish their observations and investigations, however modest or elementary they might be. They were no longer condemned to the relative isolation of small circles of acquaintance in remote provinces: *L'Astronomie* was distributed even beyond our national boundaries and came to be a link between them.

Another interest for the readers of *L'Astronomie* was that of being provided, each month, with detailed ephemerides that included, quite apart from details of the visibility of the constellations for that particular month, instructions and advice about the observation of the Sun, the Moon, the planets and all the other astronomical phenomena that were predicted (eclipses, occultations, etc.). Flammarion thus repeated there what he had been publishing regularly in the *Magasin pittoresque* since 1865, and we should note that from 1893 the *Annuaire astronomique Flammarion* was published as a separate volume. Well presented and at a modest price, this yearbook rendered an invaluable service to several generations of observers before finally disappearing in 1966.

We must also mention the planispheres, the charts and globes of the Moon and of Mars that Flammarion prepared, or had prepared from his instructions by skilled specialists such as Léon Fenet, Paul Fouché, and the selenographer Gaudibert. We may well conclude that amateurs, at the end of the last century, had all the material means at their disposal for undertaking whatever form of research interested them. This was, in large part, thanks to the example and the impetus given by Camille Flammarion.

Until 1887, however, he still lacked one other essential, the scientific background and moral encouragement that would come from a national association, which would establish a lively exchange of views between all French-speaking observers, and would give them that stimulus and pride of contributing to the progress of science.

Flammarion was aware of this. He put forward the idea for the first time in 1879 in *L'Astronomie populaire*. The success of that publication and the others that we have mentioned prompted him to gather together a group of friends who were interested in astronomy on the 28th January, 1887, and with them, to found an astronomical society. The prospectus of that first meeting clearly set out the aim of the society: "To study the best method of encouraging a love of astronomy in France, and to foster research by all those who are interested in that science."

Officially founded on 20th June 1887 and given state recognition ten years later, the Société Astronomique de France naturally made its founder the first president, and then its general secretary until his death in 1925. The office in the rue Serpente, in the scientific societies building, was, after 1890, crowned by a dome (a second was added in 1900), and there was a meridian room, a meeting room and a library. Flammarion had long dreamt of a public observatory, and as soon as it was started it became a breeding-ground for young amateurs such as Baldet, Chrétien, Danjon, Quénisset, Rougier and many more.

But this is yet another story and we have insufficient time to spend on describing a whole century that has been so full of incident. We hope that we have shown that Flammarion's was the driving force in the development of the fashionable interest that eventually became amateur astronomy. An experienced observer, who was familiar with the technical problems and the practical aspects of the discipline, after twenty years of popularizing astronomy with great literary talent and undoubted poetic verve, but purely by means of his books, he had the great merit of not shutting himself in his observatory as if in some ivory tower. He was able to see that among those whom he was educating there were many who were able – and moreover expressed a desire – to play a more active part in the great movement of scientific emancipation that occurred at the end of the 19th century. He was himself an ardent proponent of that movement – did he not have inscribed in letters of gold that profession of faith that can still be read today on the front of his observatory at Juvisy: AD VERITATEM PER SCIENTIAM?

Beginning astronomers could never have found a better guide than Camille Flammarion to the discovery of the marvels of the universe, nor a better support than that of the Société astronomique de France in encouraging them to persevere.

[A longer article on this subject has been published as: "The observatory at Juvisy", J.C. Pecker and J. Pernet, *L'Astronomie*, May 1987, p.331 – Eds.]

The Historical Contribution of Amateurs to the Study of Double Stars

P. Couteau
Observatoire de Nice, B.P. 139, F-06003 Nice, France

Double stars are favourite objects for amateurs, who can see them with modest instruments and easily observe their eventual motion. Moreover, it is possible to calculate their orbits, and even masses, with no other equipment than a pencil, a piece of thread and two pins to draw the ellipses that these systems have been following for centuries.

Binary stars are part of those classes of objects whose geometrical structure remains the same at different scales, they are internally homothetic [1] and represent a field to which fractal geometry may be applied. All this explains the popularity of these objects with amateurs: anyone is able to see some of them, whatever the power of their instruments.

For a long time astronomers were amateurs, it is only comparatively recently that they have become professionals, when administrative structures arose to regulate the profession. Men have devoted their fortunes, their lives, and their comfort, to the study of the heavens and of the laws of nature: Galileo (1564–1642) is a very good example.

William Herschel (1738–1822) was the first to draw attention to double stars as being able to demonstrate Keplerian motion. His fundamental paper on this dates from 1803, and he cites nine systems of which the orbital motion was definite.

William Herschel was not an astronomer by profession, he was a musician and organist at the Chapel Royal at Bath. He was not fettered by research, unlike the astronomers by profession who were busy with cartography concerned with the form of the Earth and with navigation. He was given a comfortable pension by King George III after his discovery of Uranus in 1781; that pension allowed him to construct his telescopes, which were by far the most powerful and advanced for that period.

In 1821 W. Herschel published two catalogues of double stars, with a total of 812 systems, many of which had been discovered by his predecessors and in particular, by Christian Mayer (1779). The way for his successors was mapped out. One of these was to be William Herschel's own son, John (1792–1871), who discovered more than five thousand double stars from the Cape and London, with telescopes built by the family.

[1] Homothetic: such that for a given set of curves, a line through any point intersects all the curves of the set at the same angle – Eds.

Russia was not long in becoming famous for its astronomical research thanks to Wilhelm Struve (1783–1864), who was given the task of constructing the imperial observatory near St Petersburg. Astronomers from the neighbouring Germany, such as Olbers, Gauss, Encke, Humboldt and Bessel helped with their advice. The largest refractor in the world, the 24-cm equatorial, built by the famous firm of Fraunhofer at Munich, was ready for service at Dorpat in 1824.

What happened was what Sir James South described as "the golden harvest of double stars from the fields of heaven". In three years, this instrument had enabled more than 1600 objects to be discovered, and in 1837 W. Struve published the famous Dorpat catalogue, which contained ten thousand measurements of three thousand pairs, a model of its kind that remained for about a century the bible of double-star observers. Later, in 1843, father and son built the fine observatory of Pulkova, near St Petersburg, with a 38-cm refractor. This observatory and the instruments it contained remained the most modern and the most powerful in the world until the 1860s, when the United States of America vigorously threw itself into the race to have the largest refractor, under the driving influence of enlightened amateurs.

Before coming to this race for large telescopes, which was to be of great benefit to the study of double stars, we must speak about two outstanding men: William Rutter Dawes (1799–1868) in Britain and Baron Ercole Dembowski (1812–1881) in Italy.

The Reverend W.R. Dawes was a medical practitioner and clergyman; his taste for astronomical observation dates from 1826, when he examined the double-star catalogues of Sir William Herschel. In 1829 he acquired a Dollond equatorial of just under 4 inches, and he made 600 measurements, which made a great impression on everyone, and he was named "the eagle-eyed". In 1853 he came into contact with Alvan Clark, who was to become the manufacturer of the largest American objectives. Clark had difficulties in starting as an optician and he sent Dawes a list of double stars, really celestial jewels, that he had discovered with small objectives that he had made. Dawes ordered three objectives of 7, 7.5 and 8.25 inches, with which he was to make some startling discoveries, whilst all the time ministering to bodies and souls, and also despite his health, which rarely left him free from trouble.

In 1850 he discovered the Crêpe Ring of Saturn, independently discovered the same year by W.C. Bond with the new 15-inch refractor at Harvard. He also produced some remarkable drawings of Mars, and saw markings on Ganymede and Callisto. He made his own micrometers, and devised ingenious diaphragms to improve this or that observation. It was he who discovered the equation giving the resolving power of an objective. Finally, he also discovered some real celestial jewels, one of which Dawes 1 = ADS 13946 is now one of the most interesting in the sky. The separation between the components is now 0.10 arcsec and is continuing to decrease from its value of 0.60 arcsec in Dawes' time.

Dawes measured almost all the objects in the Dorpat catalogue, and his measurements are of astonishing accuracy, which, having been made more than a century ago, are of the highest value for the calculation of orbits. This observer measured objects to 40% of the resolving power, with angular errors that did not amount to more than a few degrees. He was the link between the Struves and the new generation that America was about to foster.

Baron Ercole Dembowski, the son of a Polish general who served under Napoleon, lived most of his life in Italy. He was a sailor in the Austro-Hungarian service, but it would appear that the long nights at sea and the spectacle of the heavens awoke in him the wish to devote himself to the study of those distant spheres. He bought a 135-mm dialyte refractor and installed it at Cremano near Naples. Six years later he acquired a Merz 162-mm refractor and moved to Gallarate, near Lake Como, where he had an observation tower built adjoining his villa.

E. Dembowski was to make a total of 21 000 measurements over 30 years, of which 18 000 were from Gallarate. He was one of the most assiduous observers of double stars, and the most prolific of his time.

Dembowski, like Dawes a few years before, undertook the measurement of the stars in the Dorpat catalogue with a precision equal to that of Dawes. But then another amateur, perhaps the most remarkable of the three, Sherburne Wesley Burnham (1838–1921) began his remarkable discoveries, at Chicago, close to the Dearborn Observatory. This was in 1873; Burnham had only a 6-inch (15-cm) telescope, the objective was perfect, but he had neither a drive nor a micrometer. So it was Dembowski who measured the pairs discovered by Burnham, whose contribution was soon to provide a decisive spur in the drive for large telescopes.

Dembowski and Burnham worked eight thousand kilometres apart, but their observations complemented one another. Burnham was to show his gratitude to Dembowski in dedicating his famous catalogue of 1290 double stars to him in 1900.

Until 1873 the Dorpat catalogue (Σ), augmented by the Pulkova one (OΣ) formed the definitive work for known doubles. Those of John Herschel were neglected as being too wide, or else situated in the southern hemisphere. Those two catalogues provide evidence for the motion in a large number of binaries, which encourages observers. W.R. Dawes and E. Dembowski guaranteed continuity in the middle of the 19th century, and they showed that the observation of double stars is more a question of the person involved than one of equipment. The selection has been carried out, because thanks to them pairs with orbital motion have been detected, and numerous orbits are becoming visible.

In 1838 a man was born in Thetford, a small town in New England, thanks to whom, whilst never an astronomer by profession, the study of double stars was to develop into one of the major disciplines by the beginning of the 20th century: this was Sherburne Wesley Burnham. It was he who steered the drive towards the larger telescopes that revolutionized the observation of double stars.

Burnham was, by profession, an official court reporter at Chicago. He was highly regarded in his job and became a member of the Board of Administration of the Chicago and Northern Pacific Railway. His attention to astronomy was drawn by reading a copy of Burrit's *Geography of the Heavens*. His home was right next door to the old Dearborn Observatory, where the famous 18.5-inch Clark, the largest refractor in the world, had just been installed. Burnham was a frequent visitor to the establishment's library and became friendly with Alvan Clark, from whom he ordered a 6-inch refractor at a cost of eight hundred dollars. The telescope was delivered in 1873. In four years, Burnham was to discover 451 binaries, many of which are notable, among them β Delphini, with a period of 26 years.

It caused a revolution in the rather rigid world of double stars. The "golden harvest" had not been exhausted, indeed a young enthusiast, with a very modest-sized instrument had shown the contrary. And this young amateur was discovering binaries with his small refractor, whilst the fine Dearborn equatorial, the largest in the world, a few hundred metres away, had hardly been used since the terrible fire of 9th October 1871, that national catastrophe that had razed a large part of Chicago.

The press took up the cause, funds were acquired and Burnham was named Acting Director of the observatory, without a salary. He is the only amateur to be named director of a major observatory. He revitalized the establishment and went on to discover 413 new pairs there, before resigning his post because his job proved so absorbing. His friend, G.W. Hough succeeded him. But Burnham had become an important figure in American astronomy; it was his advice that decided the site of the Lick Observatory, where he stayed for four years, whilst his wife and six children lived at San José at the foot of the mountain. After the installation of the great Yerkes refractor, near Chicago, Burnham was granted its use over the week-ends, whilst for the rest of the week he was in his office at the courts.

He still found time to assemble a general catalogue of all doubles, a marvel of its kind, with full references for every object, the number of which is over 13 000.

Apart from all this, Burnham also encouraged pupils, who themselves were to become great professionals: W.J. Hussey (1866–1926) and R.G. Aitken (1864–1951). They were also to reap a rich harvest, but that is indeed another story.

In Europe, amateurs also achieved remarkable success. There was a whole series of them in Britain, where the drive for giant instruments was carried further by Newall, who, with his large 63-mm refractor dethroned the Dearborn telescope from first place, then handed on the laurels to Lord Rosse, who constructed his great 1.8-m reflector.

Double stars have made a few European amateurs famous, the first that we shall mention being Camille Flammarion (1842–1924). In 1868 he was the first to publish a list of pairs where orbital motion had been confirmed, and which at that time numbered eighteen. Flammarion, by his various publications, and his enthusiasm, where his romantic view sometimes gained the upper hand over the constraints of scientific rigour, contributed enormously – as did Arago – to the popularization of astronomy and the encouragement of people to take it up as a vocation.

Three other names are pre-eminent as regards double stars: R. Jonckheere (1889–1974), the Reverend T.E. Espin (1858–1934) and Dr Paul Baize, who is still working in his native Normandy.

Robert Jonckheere had a private observatory built at Hem, near Lille. His 33-cm refractor enabled him to discover 1000 pairs at a time when Aitken was carrying out his great searches; a number that he later increased to 3300. The Reverend Espin used reflectors, and around the 1920s discovered about 3000 binaries under English skies, which obviously are not always foggy. But the Jonckheere and Espin pairs have components that are too widely separated to be dynamically interesting, but their searches are remarkable for the number of objects that they discovered.

Dr Paul Baize, born in 1901, is the prototype amateur, who is both an observer and binary-star theoretician. The son of a doctor, his father naturally steered him towards a medical career, a career that has been very full, because apart from being a consultant paediatrician, he also served at the Hospice des Enfants-Assistés, as lecturer at the School of Paediatric Nursing, and as a school doctor, which was more than enough to occupy a single person. But that was not enough for Dr Baize. In 1925 he acquired a 108-mm refractor, then he had the use of the equatorials at the Paris Observatory, one a 30.5-cm and then a 38-cm. The total number of measurements made by Dr Baize is 24 044, and he is among the top ten "measurers" in the history of double stars.

The study of double stars has often been considered as of marginal interest for professional astronomers working in fundamental astronomy, and later in the development of astrophysical methods, and later still in cosmology. It was Burnham who started professionals studying binaries after the great Lick refractor was brought into service. But they have never been numerous, and it has been amateurs who have ensured continuity.

Nowadays, modern interferometric methods require complicated physics and information technology that are out of the reach of amateurs. But they are the ones who will have to observe the numerous slow-moving pairs that will only reveal their orbit much later in the future. The giant refractors are gradually being retired from scientific use; would it not be a good idea for amateurs to form groups and make arrangements to keep them maintained, as has already started to be done in certain organisations?

Then, thanks to the efforts of those who love the heavens, orbits will abound, and thanks to the Hipparchos satellite, which will provide the parallaxes, knowledge of stellar evolution will be the final beneficiary.

Bibliography – Limited to Fundamental Works

Aitken, R.G., *New General Catalogue of Double Stars*, Vols I & II, Carnegie Institute, Washington D.C., 1938: 17 180 pairs with measurements and references

Burnham, S.W., *A General Catalogue of Double Stars*, Parts I & II, Carnegie Institute, Washington D.C., 1906: 13 665 pairs with measurements and references

Dawes, W.R., his observations are gathered together in *Mem. Roy. astron. Soc.*, **5** (1833), **8** (1836), **19** (1851) and **35** (1867)

Dembowski, E., *Misure Micrometriche de Stella Doppie e Multiple*, 2 Vols, Reale Academia Lincei, 1883 & 1884

Herschel, Sir J.F.W., "A Catalogue of 10 300 Multiple and Double Stars", *Mem. Roy. astron. Soc.*, **40** 1874); covers the discoveries made by both the Herschels father and son, but is just an index and contains no epochs or measurements

Jeffers, H.M., van den Bos, W.H. & Greeby, F.M., *Index Catalogue of Visual Double Stars*, 1961: list of more than 60 000 pairs from pole to pole

Jonckheere, R., *Catalogue and measures of Double Stars*, *Mem. Roy. astron. Soc.*, **51**, 1916

Schiaparelli, G.V., *Misure di Stella Doppie*, 2 Vols, Brera, Milan, 1888

Struve, F.G.W., *Stellarum duplicum et multiplicum mensurae micrometricae*, Dorpat Observatory, 1837: the famous Dorpat catalogue

Twentieth-Century Amateur Astronomy

A.J. Meadows
Department of Library and Information Studies, Loughborough University, Loughborough LE11 3TU, United Kingdom

Little distinction was drawn between amateur and professional astronomers for much of the nineteenth century. They mixed in the same scientific societies and often carried out overlapping studies. Towards the end of the century, however, new factors arose – increasing expense of instrumentation, increasing sophistication of theoretical knowledge, etc., which led to a greater degree of differentiation. It was then that societies specifically aimed at amateurs were established. The split has never been complete – professionals have always been members of amateur societies and vice versa – but a gap between amateur and professional opened up and has continued since. Amateurs of the standing of Percival Lowell, who could compete with professionals both in terms of equipment and theoretical knowledge, effectively died out before the mid-twentieth century.

Most countries with an astronomical tradition have developed amateur societies at two levels – the national and the local (or regional). Many amateurs belong to both types: virtually all serious amateur astronomers are affiliated to one or the other. In total, the number of amateur astronomers in the world has always exceeded the number of professionals; so their activities must always be taken into account in discussing the history of astronomy – even its recent history.

The membership of amateur societies typically shows a fairly even spread of ages (with more younger members than professional societies). This is illustrated in Table 1 (Fisher, 1980). In addition, the proportion of female members is usually higher in amateur societies. The impetus for the separation of the British Astronomical Association (BAA) from the Royal Astronomical Society (RAS) was, in part, connected with this. To quote the BAA's original statement of aims, it was intended "to meet the wishes of those who find the subscription of the Royal Astronomical Society too high, or its papers too advanced; or, who are, as in the case of ladies, practically excluded from becoming Fellows". The BAA practised what it preached.

Table 1

Age in years	Under 25	25–44	45–64	Over 64
% of members	20	35	35	10

The main group of observers sent by the BAA to the solar eclipse of 1898 consisted of ten men and five women. Nevertheless, the proportion of women in amateur astronomical societies – typically 10–20% – has always been considerably lower than in amateur natural history societies.

Amateur astronomical societies are seen much more than professional societies as a means of obtaining training. Stebbins (1980) has distinguished between three types of amateur astronomer – the apprentice, the journeyman and the master – in a way that emphasises this training aspect. The apprentice is primarily concerned with learning about the subject; the journeyman is a knowledgeable practitioner who can work independently; the master is actually in a position to contribute to astronomy. The journeymen turn to the masters when they run into difficulties. Both groups pass on their knowledge and experience to the apprentices. The importance of this training activity stands out clearly when amateurs are asked what the chief aims of their society should be – see Table 2 (Fisher, 1980).

Table 2

Essential aim	% fully agree	% partly agree
To provide a meeting place for like-minded people	90	7
To provide more theoretical knowledge	70	23
To provide facilities for practical work	68	25
To provide a library	36	39

Table 2 seems to suggest that members of amateur societies place theoretical knowledge and practical work on a par with each other. In fact, this hides an interesting difference between amateur and professional astronomers. Amateurs tend to despise "armchair" astronomers: society members are expected to be out there observing or developing instrumentation, or, at least, calculating orbits. "Masters" are typically members with particular skills in these areas. In professional astronomy, on the contrary, observers often believe that they are allotted less prestige than theoreticians. Amateurs often join societies especially to gain advice about equipment. This frequently includes tuition in the building of telescopes. Earlier in this century, societies might be devoted almost entirely to telescope making. Though most now have a more balanced approach, there are still many amateurs who find working on instrumentation more satisfying than actual observation.

As Table 1 suggests, amateur societies contain appreciable numbers of young members who are still receiving full-time education. These are almost always in the apprentice category. What they learn in the society motivates a significant proportion of them to contemplate acquiring professional qualifications. Though only a small fraction actually do so, the amateur societies provide a valuable source of professional observational astronomers.

Most amateur societies try to provide some kind of publication, even if only a brief newsletter. Since this often depends on individual initiative in local societies, such publications tend to be ephemeral. At the national society level, of course, the publications are much more stable and durable. An appreciable proportion of society members (10–20%) have talked, or written about astronomy for groups outside their society. Amateur astronomers have always provided a useful channel for transmitting an idea of astronomy to the general public. Indeed, they are often as successful as professionals in presenting astronomy via the media.

Throughout most of the twentieth century, amateurs have differed from professionals in the celestial objects they preferentially choose to study. By the latter part of the nineteenth century, amateurs were finding the professional emphasis on precise positional measurement coupled with extensive mathematical analysis both too demanding and too boring for them. The new science of astrophysics attracted them immediately, and many of the early advances in this field were made by amateurs (e.g. Sir William Huggins and his wife). G.E. Hale stood at the end of this particular line of development. His early work on the spectroheliograph was carried out on an amateur basis at his own observatory in Chicago. But soon, as a professional astronomer, he was seeking to improve the provision of telescopes in the USA. This led in a few years to the demand for bigger telescopes. Increasingly, professional astronomers were changing their focus of interest to faint objects – stars and galaxies. Correspondingly, their interest in Solar-System objects diminished. Amateurs could afford neither large telescopes, nor the sophisticated ancillary equipment that accompanied them. So a division grew between the astronomical topics of amateur concern and those of professional astronomers. A professional astronomer, H.H. Turner, wrote a song for the BAA in the 1920s (to be sung to the tune of "The British Grenadiers"). One verse lists the observational interests of BAA members (JBAA, 1930).

"It [the BAA] groups itself in Sections, for Meteors or for Mars,
While one has predilections for Variable Stars,
Nor is the Moon neglected, nor Comets, nor the Sun
By the British Astronomical Association"

The separation between this amateur interest in the Solar System and the professional emphasis on stellar and galactic studies was underlined in 1930 by the discovery of Pluto, which can be regarded, from one viewpoint, as the greatest contribution of the amateur tradition to Solar-System astronomy in the twentieth century. The Lowell Observatory was still faithfully following Percival Lowell's precepts, and the planet was itself discovered by a young amateur, Clyde Tombaugh.

Over the past quarter of a century, the growth of space research has done a great deal to reduce this separation. In the early days of the space age, professionals discovered that they could learn from amateur astronomers – for example, of activity in Jupiter's atmosphere. In several areas, amateur and professional activities complement each other. An example is the discovery and observation of comets, where amateurs still play an important part despite the greatly increased professional involvement. In a quite different area of space science, the work of an amateur, H.H. Nininger, in collecting meteorite samples, proved of considerable value to professional meteorite studies in recent years. Amateurs, perhaps because their reputations were less at stake, were often more ready than professionals to consider the possibility of spaceflight in the pre-space age days. It has turned out that artificial satellite observation is probably the most important new area of amateur observation in the twentieth century. Visual observation of satellites by amateurs still provide a useful input of information to professional studies.

The division between the Solar System for amateurs and the remainder for professionals was, of course, never anywhere near complete. Some amateurs during the twentieth century have contributed to observations of visual binaries, but this is an are where professional interest has declined from the nineteenth century. The topic which both amateurs and professionals have found fascinating throughout the century is variable stars. The extent of the amateur commitment to variable-star observations is indicated by the history of the American Association of Variable Star Observers (AAVSO). This celebrated its 75th anniversary in 1986, when it had 1,300 members spread over more than forty countries. Between them these members are making annually a quarter of a million estimates of stellar brightness.

However, the most important amateur contribution to observations outside the Solar System came, not from a society, but from an isolated observer. During the 1930s, an American radio engineer, Grote Reber, followed up Jansky's initial observations of radio noise from space. He constructed the first parabolic radio antenna and used it to map the intensity of radio emission from the Milky Way. His discussions with professional astronomers and subsequent publication of his results in the *Astrophysical Journal* became one of the significant starting points for the new science of radio astronomy.

Though much has changed during the twentieth century, the role of amateur astronomical societies has not altered very greatly. *L'Astronomie* (1937), reporting on the fiftieth anniversary of the Société, had this to say:

> "Un de nos collègues nous demandait un jour le secret du succès de la Société Astronomique de France et de l'atmosphere si sympathique qui y règne. Ce secret, nous semble-t-il, est que toutes les fonctions y'étant gratuites et les collaborations bénévoles."

This description surely has a wider applicability in amateur astronomy. A few years ago, young astronomers went round in T-shirts bearing the inscription – "Astronomers do it at night". I translate this passage from *L'Astronomie* as an equivalent inscription applicable to all ages – "Astronomers do it with pleasure".

References

T. Fisher (1980). *The Role of the Amateur in Science*, Ph.D. Thesis, University of Leicester, U.K.
JBAA (1930). *Journal of the British Astronomical Association*, **41**, 52
L'Astronomie (1937). *L'Astronomie*, **51**, 503
R.A. Stebbins (1980). *International Journal of Comparative Sociology*, **22**, 34

Criteria for Identifying an Astronomer as an Amateur

Thomas R. Williams
3621 Wakeforest, Houston, TX 77098-5509, U.S.A.

While significant contributions of amateur astronomers are generally recognized in the literature, the identification of the individual as an amateur is frequently unclear. As a result, amateur astronomers today have a limited sense of the contributions of earlier amateur astronomers. In part, this problem stems from current usage of the word "amateurish" as a pejorative, representing something not well-done or lacking in quality. In addition, many individuals, who take pride in identifying themselves as amateur astronomers are neither inclined, nor in fact capable, of making a contribution to astronomy. It is important, however, for historians and others who write about the history of astronomy to recognize amateur astronomers and identify the significance of their contributions. For it is through such elaboration that other amateurs will recognize the possibilities and great value that may be associated with their own efforts. Therefore it is necessary to develop a common understanding of the characteristics that distinguish an amateur astronomer from what I identify in this paper as a "recreational sky observer". Such a classification scheme should help future writers to correctly identify amateur astronomers and their contributions.

In considering the term "amateur astronomer", I focus on the fact that one must first qualify as an astronomer. Only then is it possible to consider further classification as an amateur or a professional. From dictionary definitions the following statement may be derived: An astronomer is a person who is very skillful or highly trained and informed in that branch of knowledge concerned with establishing and systematizing facts, principles and methods, and with conduction experiments and observations in order to develop hypotheses and systematized knowledge about the nature or principles of the stars, planets and all other heavenly bodies, dealing with their composition, motion, relative position, size and other properties.

With this definition in mind several criteria may be developed to classify an individual as an astronomer and to thereby differentiate him from a recreational observer: (1) A serious intent must be displayed to contribute to the advancement of astronomy by providing information that is needed and might otherwise not be available to other astronomers; (2) These efforts must extend over a period of time, either by routine observations, or through discovery or search work which can be either theoretical or observational; (3) Acceptable methods or techniques must be used considering the intended application of the data; and (4) The results of this work should be communicated to other astronomers.

Consideration may now be given to characteristics which differentiate professionals and amateurs. A professional is a person who engages in some art, science, sport, etc. for his livelihood. An amateur is a person who engages in some art, science, sport, etc. for the pleasure of it rather than for money. The question of whether one is paid or not becomes the critical issue in differentiating between the professional and the amateur. The amateur is likely to derive his income from other means than astronomy. Most amateur astronomers lack the theoretical training of the professional. Without that common language, it is difficult for them to communicate about the questions that need answering, even though the answers may be within the amateur's grasp. Yet amateur astronomers frequently excel in observational astronomy and have much to offer in terms of their dedication to extensive routine observation.

While recreational sky observers appreciate the beauty of the night sky, they are not motivated to contribute to the science. Their skills may be significant in locating and identifying faint or small objects, even occasionally a comet or nova. However, their routine usually includes re-observation of the Messier Catalogue or checking the colors of bright double stars. "Armchair" astronomers who take pleasure in reading everything available about astronomy are also included in this group.

The key to differentiating between recreational sky observers and astronomers, whether amateur or professional, is in observing that astronomy is work, that certain features of this work fit a recognizable pattern constituting the practice of astronomy, and that individuals who do not work within these recognizable patterns should not be considered astronomers, amateur or otherwise. These criteria should facilitate the identification of amateur astronomers, who attempt to make significant contributions to astronomy. Identification of amateur astronomers will restore an appreciation of the past contributions made by amateur astronomers, and will encourage current and future efforts by amateurs for the advancement of astronomy.

Reference

Williams, T.R.: *Criteria for selection of an astronomers as an amateur*, Abstract 51.11, Bull. Amer. astron. Soc., **16** (2), 1984

Johannes Hevelius: Polish Seventeenth-Century Brewer and Astronomer

Cecylia Iwaniszewska
Institute of Astronomy, N. Copernicus University,
P-87-100 Torun, Poland

On January 28th, 1987 Polish astronomers commemorated the tercentenary of the death of Johannes Hevelius. Born in 1611 of a rich bourgeois family of Gdansk, a Polish merchant city by the Baltic Sea, he studied law at Leiden. But he soon became more interested in sciences and travelled to London; to Paris, where he knew Mersenne, Gassendi and Boulliaud; to Switzerland; and to Germany. On returning home to Gdansk he had to take over the administration of family properties : breweries, stables, town-houses, but he got considerable help from his wife Catherine Rebeschke whom he married in 1635.

He could then turn his attention entirely to observing celestial bodies, his first observation of an annular solar eclipse made him aware of the poor quality of existing lunar charts. So he devoted himself to systematic observation of our satellite, observing the Moon with home-made telescopes, and himself engraving his drawings of lunar details. The instruments installed on a terrace upon the roofs of three adjoining houses formed his observatory, called STELLABURGUM – City of Stars.

On hearing that Gassendi in Paris had also prepared a moon map, Hevelius sent him a sample of lunar drawings and asked for his opinion. These drawings must have been of a very superior quality, because Gassendi reply enthusiastically :"I owe you many thanks for the beautiful drawings (...). You are gifted with such superior eyes that the representation could not be improved upon (...). I urge you to bring to light the Moon description that I had planned". The Moon description was been published in Gdansk in 1647 as "Selenographia" (Selene = Greek goddess of the Moon). The volume contained forty lunar charts at different phases, with details, names, and mountain heights with Hevelius' own method. For over 150 years, the book was the best lunar atlas and had a very high reputation among scientists in France, England and Italy.

Hevelius wanted to follow this with another project – the preparation of an extensive catalogue of star positions. For these observations he prepared a whole set of instruments : quadrants, sextants and octants, solidly made out of wood or brass and richly adorned. His favourites were a 5-ft quadrant and a 6-ft sextant with plain sights. The first part of the catalogue describing methods and instruments was published in 1673 as "Machina Coelestis" part I; the second part of "Machina Coelestis" decribing observations, appeared in 1679, while the "Catalogus Stellarum Fixarum" with 1888 star positions was been published only in 1690, after Hevelius' death. He also observed the Sun and planets, decribed and discovered comets, publishing the results in his own printing office after he obtained the necessary permission from the Polish king Jan II Casimir, in 1659. In 1664 he was elected as the one hundred and forty-seventh fellow of the Royal Society of London. After the death of his wife Catherine in 1662, he married, in 1663, Elisabeth Koopman, more than thirty years his junior, who not only took care of the house and family, but acted as faithful assistant to Johannes.

One event disturbed the peaceful years of observation and work : the publication in 1674 of a critique of the accuracy of Hevelius' observations by Robert Hooke. Although the observations of Flamsteed at Greenwich in 1676 (with sextants equipped with telesopes) seemed to agree with those of Hevelius, the Royal Society found it best to send young Edmond Halley to Gdansk in 1679 to examine the local instruments and observations made with them. Halley's tests proved the accuracy and reliability of Gdansk measurements, which was duly reported in the "Royal Society Philosophical Transactions" : "Mr. Hally (!) declares himself abundantly satisfied of the use and certainty of these instruments and observations (...) having seen with his own eyes a multitude of observations (...) with the great brass sextant, again and again repeated, most accurately, and almost incredibly to agree."

A few weeks after Halley's visit a fire destroyed STELLABURGUM; houses, instruments, library, printing office, all was gone; only the unfinished catalogue manuscript was saved. Johannes Hevelius was already aged 68, but nevertheless he undertook the building of a new observatory. He got the necessary funding from the Polish king Jan III Sobieski, and the French king Louis XIV, bought new instruments (which proved to be of much poorer quality), and finally completed the star catalogue, but did not live long enough to see it printed.

The final volume was printed by Elisabeth together with "Firmamentum Sobiescianum" – an atlas of stars dedicated to king Jan III Sobieski. Hevelius had been preparing this atlas for many years and introduced twelve new constellations, one of them being "Scutum Sobiescianum" – the shield of Sobieski with the coat of arms of the King. The constellation "Scutum" still remains on all sky charts, only the name of the Polish king has been omitted. Another constellation named by Hevelius is "Lynx", so named because it consists of very faint stars that can only be seen by a person having the eyes of a lynx, as was the case with Hevelius himself. Yet another constellation is the "Sextant", named after Hevelius' favourite instrument.

Today very few astronomers seem to know that the great observer Tycho Brahe had a worthy successor, and that there was a Polish brewer on the Baltic Sea who owned the greatest European observatory in the mid-seventeenth century, even before the Paris and Greenwich Observatories came into existence, in 1667 and 1675 respectively.

How could this remarkable Moon observer have foreseen that on the 299th anniversary of his death, astronomers would obtain direct pictures of a far- away planet, and pictures of surface details of the satellites of Uranus, among them of little Miranda, many times smaller than the terrestrial Moon...

Eugène Antoniadi (1870–1944) and Planetary Observation

Adouin Dollfus
Observatoire de Paris, Section d'Astrophysique, 5, Place Jules Janssen, F-92195 Meudon Principal Cedex, France

Born a Greek citizen at Constantinople on 1870 March 10, Eugène Michel Antoniadi was a cultured man, well-read and scholarly, an historian and linguist, but above all an artist. His artistic tools were pencils, washes, pastels and watercolours, all handled with exquisite delicacy. This talent was the all-important factor in his magnificent contribution to observational planetary astronomy.

In 1888, at the age of 18, he made his first observations, using a small 75-mm refractor, which was soon replaced by a 108-mm instrument. He drew the appearance of the sky and of the planets and sent his observations to the great Camille Flammarion in France. Some of them can be found in the earliest numbers of *L'Astronomie*.

He was a great admirer of French culture and, intrigued by Camille Flammarion's striking personality, he decided, at the age of 23, to come to Paris. That was in 1893. Flammarion welcomed him at Juvisy Observatory. Access to the 242-mm refractor provided his artistic talent with unparalleled views of the planets. The first books of observations – a marvellous series of pictures, like medieval illuminations – are carefully preserved at Juvisy, entitled "Juvisy Observatory: Observations made under the direction of Camille Flammarion by Eugène M. Antoniadi". Volume I, dated 1896, contains some exquisite drawings, watercolours and wash-drawings of the nebulae in Orion and Lyra, the pair Messier 65 and 66, the occultation of Jupiter by the Moon on 1896 June 14, the zodiacal light, the Moon in front of the Pleiades, Venus, Jupiter, Saturn and, in particular, Mars, for which he had already produced a projected chart.

There was to be a total eclipse of the Sun on 1896 August 8, when the zone of totality would cross the Arctic Ocean, just touching North Cape. Camille Flammarion decided to entrust the young Eugène Antoniadi with the scientific expedition. This was a considerable journey for those days, and although the sky was covered at the time of the event, Antoniadi brought back exquisite watercolours and wash-drawings for the Juvisy records.

Then, at Juvisy, Antoniadi deliberately turned towards the study of the planets. His portefolios accumulated drawings in true colour of Jupiter, Saturn and Mars, whilst his wide cultural interests were employed to publish his discoveries. This was the period when Percival Lowell, on the other side of the Atlantic, was publishing his strange works on Mars. The "canal" theory was rampant, whilst surface markings seen through the telescope revealed changes linked with the seasons. Antoniadi set about clarifying the problem and the study of Mars was to remain his principal preoccupation for the rest of his life. The British Astronomical Association soon entrusted him with the Directorship of its Mars Section.

At the same time, E.M. Antoniadi undertook, in 1904, a fundamental, historical work, a major publication, the *Atlas of the Mosque of Saint Sophia at Constantinople*. Published in 1907 in three large volumes, in Greek, this extremely rare work documented, for the first time, the interior of this famous basilica, which was turned into a mosque after the fall of Constantinople. More than 1000 photographs and innumerable watercolours, drawings and plans are reproduced and documented. A unique feature was that Sultan Abd-ul-Hamid gave special authority for the photography of the interior of the famous mosque for the first time.

In 1909, an important event again helped to decide the direction of his work. That year Mars was closest to the Earth. Henri Deslandres, Director of the Meudon Observatory gave Antoniadi permission to use the great 83-cm refractor, which was the most powerful instrument then available in Europe, and almost in the world. This was a revelation. Taking advantage of atmospheric conditions that are rarely so favourable, Antoniadi had a view of the planet that surpassed everything that anyone had ever had previously. Moreover, he had the talent to reproduce what he saw with his pencil or with his brush. He was being offered the chance of taking planetary exploration a stage farther, and he understood that. Calling himself "an honorary astronomer", he dedicated most of his later activity over a period of more than thirty years to that end.

With Mars he saw the myth of the "canals" disappear in front of his eyes. In 1897 he may have written in the Juvisy observing books, as a conclusion of his observational analyses "We cannot avoid concluding that the famous canals of Mars do have a truly objective existence", but he promptly revised his judgement. In 1930, his vigorous pen firmly stated:

"No one has ever seen a true canal on Mars, and therefore the essentially straight, single or double "canals" of Schiaparelli's do not exist as canals, or in any geometrical form. They do, however, have a basis in reality, because at the site of each of them, the surface of the planet shows either an irregular streak that is more or less continuous and speckled, or a ragged, greyish edge, or even an isolated, complex lake."

Antoniadi described the seasonal variation of the martian polar caps. He studied the short-lived clouds that are seen occasionally in the atmosphere, and well as the dust storms that sometimes obscure vast areas. He also depicted, as no one had before, the evolution of the spots and strange marblings detected on the surface of the planet, and his artistic talent allowed him to record colour variations. His scholarship enable him to trace the history of the seasonal or random variations over more than a century. His hellenic, classical and mythological backgrounds helped him in the problem of nomenclature, because the markings on the martian surface are identified with names borrowed from the Graeco-latin world of the Mediterranean. They are extremely evocative. Antoniadi introduced both order and harmony. As a result, his comprehensive publications *The Planet Mars*, a monumental work dating from 1930, remains, more that fifty years later, the bible for the telescopic mapping of Mars.

As regards Jupiter, through the refractor Antoniadi saw a dense, turbulent and restless atmosphere. Against the astrophysical evidence he argued, with his acid pen, for a very hot planet, essential, he though to account for the eddies that he saw and drew. Here, the artist gained the upper hand on the physicist. On Mercury, his observations enabled him to make out a characteristic feature in the shape of the numeral 5. The recurrence of this reference point led him to announce that its rotation was about an axis perpendicular to the orbital plane and had a period of 88 days, equal to the planet's orbital period around the Sun. To him the situation seemed similar to the case of the Moon and the Earth, and being accounted for by tidal theory. Antoniadi showed that the tidal forces producing synchronous rotation would vary as the 6th power of the distance from the central body. His work *The Planet Mercury*, published in Paris in 1934, asserts the reality of the 88-day rotation in very definite terms.

We now know that the rotation of Mercury is 2/3 of that period, i.e. 58.6 days, and that this exact fraction results from the combination of tidal forces and the eccentricity of the orbit. The progress of science sometimes takes place like this, Antoniadi was the victim of a stroboscopic effect between the true rotation period and that of elongations easily observable from Earth, and he buttressed his arguments with an only approximate theory.

The linguistic and hellenistic talents that inspired Antoniadi also led him to an important work, *Egyptian Astronomy*, published in Paris in 1934. In this, Antoniadi examines and describes the astronomical knowledge of ancient Egypt, and of the Greek and Mediterranean world. In a striking parallel, he compares extracts from Copernicus' *De Revolutionibus Orbium Coelestium* with ancient texts. He shows that Copernicus drew the heliocentric idea that made his fame from his knowledge of Greek works.

Apart from the publications mentioned, all of which were of fundamental importance, Eugène Antoniadi wrote more than 40 astronomical papers, many of which appeared in *L'Astronomie*, the journal of the Société Astronomique de France.

Naturalized French in 1928, he was later honoured by being made a Chevalier of the Legion d'Honneur. The Académie des Sciences awarded him the Prix Lacaille in 1932, then the Prix Gusinan in 1940. The great "honorary astronomer" also received from the Société Astronomique de France the prestigious Prix Janssen, which honours a major contribution to astronomical knowledge. This was a unique award to an amateur. Eugène Antoniadi died in Paris on 1944 February 10, in his beloved France, which was in the middle of a war and occupied by the enemy.

Publications by Antoniadi

Books

La planète Mars, Hermann, Paris, 1930
La planète Mercure et la rotation des satellites, Hermann, Paris, 1934
L'astronomie egyptienne, Gautier Villars, Paris, 1934

Articles

"Le retour de la planète Mars", *L'Astronomie*, 1926 Aug. p.346
"Les satellites de Jupiter", *L'Astronomie*, 1927 Aug. p.353
"Le retour de la planète Mars", *L'Astronomie*, 1928 Dec. p.564
"La rotation des satellites", *L'Astronomie*, 1929 Sept. p.385
"La planète Saturne", *L'Astronomie*, 1930 Feb. p.49
"La planète Mars (1659–1929)", *L'Astronomie*, 1930 Sept. p.411
"La planète Mercure", *L'Astronomie*, 1933 Dec. p.545

Investigating Astronomy's History

Norman Sperling
4917 Mountain Boulevard, Oakland, CA 94691, U.S.A.

Abstract

Amateur astronomers play an important role in historical astronomy and publish a wide range of studies. They can participate nearly as well as professional astronomers since neither are professional historians or archaeologists. Amateurs often restore old instruments and facilities, discovering clues to their design, manufacture and use that would not be noticed by non-astronomical historians.

Possible studies include histories of institutions and of astronomy in different countries, biographies, and evolution of ideas. Publications and manuscripts could be studied for matters of current and historical importance. Folklore (of all countries) should be recorded before it dies out, and the beliefs behind it examined. More study groups involving amateurs, professionals and others should be established.

Charles Boyer and the Rotation of Venus

Adouin Dollfus
Observatoire de Paris, Section d'Astrophysique, 5, place Jules Janssen, F-92195 Meudon Principal Cedex, France

In August 1957, the magistrate Charles Boyer, who was President of the Court of Appeal at Brazzaville in Africa and an amateur astronomer, began to photograph the planet Venus systematically, using a violet filter and his own personal telescope, 256-mm in diameter, with an optical window. His aim was to study the positions of the dark, changing markings that had been detected in ultraviolet light by Ross at Mount Wilson in 1924. He was collaborating with Henri Camichel at the Pic-du-Midi Observatory, who was simultaneously observing with a 60-cm reflector. Over 68 days' observation, Charles Boyer noted a periodicity of 4 days in the appearance of the markings. Alerted by this, Henri Camichel found the same effect on his series of images. Charles Boyer photographically added together the images on plates obtained at daily, 3-day, 4-day, and 5-day intervals (1). The combination of different images cause the markings to disappear. Only superimposition of 4-day images revealed any detail, the result of periodicity in the planet's appearance.

From new observational campaigns undertaken from Brazzaville in 1959 and 1960, as well as the examination of older photographs taken at the Pic-du-Midi by Audoin Dollfus in 1948 and Henri Camichel in 1953–4, Charles Boyer was able to detect an alternation of central and limb markings that resembled a horizontal "Y", reappearing every 4 days (2). The motion seemed to be from East to West, suggesting that there was a 4-day, retrograde rotation about an axis that was essentially perpendicular to the orbital plane. The intersection of the arms of the Y could be taken as the origin of a system of coordinates. Charles Boyer calculated ephemerides using a period of 96^h33^m. Later observations showed a certain variation about this mean period (3, 7).

In order that the appearance of these cloud markings could be followed without the interruptions caused by the planet's rising and setting, Meudon Observatory organised a world-wide campaign for the photographic observation of Venus in the ultraviolet, beginning in 1962, and to which 8 stations contributed. These were located at different terrestrial longitudes in order to obtain images following one another at intervals of a few hours (4, 5). Between 1964 and 1966, Bernard Guinot and Martine Feissel, who examined the Doppler effect by spectroscopic interferometry, independently found a retrograde rotation with a period of -4.3 ± 0.4 days. Yet Charles Boyer's results have continued to meet with great scepticism outside France.

Charles Boyer and Pierre Guérin succeeded, however, using the 105-cm telescope at Pic-du-Midi, in photographing the markings in ultraviolet, in full daylight, and in following their motion from East to West more or less directly, and several times for 6 hours (6).

They were able to draw up a chart of the average appearance of the recurrent features in the cloud layers, and where the horizontal Y is related to a similar feature like the Greek character psi, turned through 90°.

The periodic return of the equatorial Y and psi features again confirms the average period of retrograde rotation of these markings as being almost equal to 4 days (7, 8). This period, which applies at the equator, increases noticeably when the latitude of markings exceeds $\pm 6°$ (8). The velocity of the horizontal displacement is found to be slower, $83\,\mathrm{ms}^{-1}$, in the morning before the subsolar point is reached, but to accererate afterwards, reaching $122\,\mathrm{ms}^{-1}$ in the afternoon (9, 10).

In 1974, more or less all the photographs of Venus taken in the ultraviolet since the discovery of the UV markings by Ross in 1927 that were available in the world were brought together at the International Astronomical Union's Planetary Photography Centre by Charles Boyer and Audoin Dollfus. These were subjected to systematic analysis (11). For 1972, in particular, the numerous, excellent photographs from the Lowell, New Mexico, and Table Mountain Observatories and, above all, from Pic-du-Midi – especially those obtained by Charles Boyer and Michel Aurière – enabled practically continuous surveillance of the appearance of the clouds over 28 consecutive days. This coverage was shown by 7 charts covering as many consecutive rotations of Venus (11).

In the meantime, the Soviet spaceprobe Venera 8 had landed a capsule on the surface of Venus on 1972 July 22, and VLBI (Very Long Baseline Interferometry) analysis of the signal emited during the parachute descent had enabled V.V. Kerzhanovich to detect a retrograde atmospheric flow of $100\,\mathrm{ms}^{-1}$ at a height of 55 km. On its journey to Mercury, the American spaceprobe Mariner 10 flew past Venus on 1974 February 5, carrying on board a telescope specially fitted with an ultraviolet filter, and returned a regular sequence of images during the approach phase. These were reconstructed into the form of a movie film, which immediately showed that the clouds were rotating in the retrograde sense discovered by Charles Boyer at an average velocity of $-110\,\mathrm{ms}^{-1}$. The Y and psi cloud features could be seen distinctly.

Profiting from all these results, Jacques Blamont proposed releasing balloons into the atmosphere of Venus, which, carried by the winds, would be able to analyze the atmospheric properties over a long track. This proposal was first studied by Michaël Marov, when it was known as Project EOS, and it finally materialised in 1985 as part of the Soviet Vega mission. On 1985 June 10 and June 14, 2 balloons were released from an entry capsule into the atmosphere of Venus, inflated whilst hanging from a parachute, and stabilized at 50 km altitude. Each flew at this level for nearly 2 days, and covered half of the equatorial circumference of Venus, carried from East to West at $100\,\mathrm{ms}^{-1}$ by the atmospheric flow discovered by Charles Boyer.

References

1. Boyer, C. & Camichel, H. (1960), Société Astronomique de France, Commission des Surfaces Planétaires, "Photographic study of Venus in violet and ultraviolet light", [in French], *L'Astronomie*, **74**, 375–381
2. Boyer, C. & Camichel, H. (1961), "Photographic observations of the planet Venus", [in French], *Ann. Astrophys.*, **24**, 531–535
3. Boyer, C. & Camichel, H. (1965), "Photographic study of the rotation of Venus", [in French], *C.R. Acad. Sci. Paris*, **260**, 809–810
4. Boyer, C. (1965), "Study of the rotation of Venus", [in French], *L'Astronomie*, **79**, 223–228
5. Dollfus, A. (1968), "Synthesis on the ultra-violet survey of clouds in Venus' atmosphere", in *The Atmospheres of Venus and Mars*, Gordon and Breach, London, pp. 133–145
6. Boyer, C. & Guérin, P. (1966), "Direct evidence by photography of a retrograde 4-day rotation of Venus", [in French], *C.R. Acad. Sci. Paris*, **263**, 253–255
7. Boyer, C. & Camichel, H. (1967), "Determination of the rotation velocity of markings on Venus", [in French], *C.R. Acad. Sci. Paris*, **264**, 990–992
8. Boyer, C. & Guérin, P. (1969), "Study of the 4-day retrograde rotation of the upper atmospheric layer of Venus", [in French], *Icarus*, **11**, 338–355
9. Boyer, C. & Guérin, P. (1971), "Irregularities in the 4-day rotation of cloud markings in the shape of a Y in the upper atmosphere of Venus, and their correlation with the slow rotation of the surface", [in French], *C.R. Acad. Sci. Paris*, **273**, 154–157
10. Boyer, C. (1973), "The 4-day rotation of the upper atmosphere of Venus", *Planet. Space Sci.*, **21**, 1559–1561
11. Dollfus, A. (1975), "Venus: Evolution of the Upper Atmospheric Clouds", *J. Atmospheric Sci.*, **32**, 1060–1070

The Amateur and Eclipsing Binary Stars

Frank Bradshaw Wood
Dept. of Astronomy, University of Florida, Gainesville, FA 32611, U.S.A.

Probably no other branch of science has benefited as much from the work of dedicated amateurs as has the science of Astronomy. While some amateurs have made many useful types of astronomical observations— comets, meteors, occultations, etc. — perhaps none has been as extensive and as useful as those made in the field of variable stars. There are not nearly enough professional astronomers to keep under proper observation the increasingly large number of known variables. While all kinds of them are well worth continuous study, this paper will call attention to the importance of eclipsing variables and in particular the systematic and continuing observation of their times of minimum light.

The systematic study of eclipsing binaries goes back to the year 1783 and an amazing amateur astronomer, John Goodricke. His paper in the Philosophical Transactions of the Royal Society, London (Vol. 73, p. 474) not only announced the variability in brightness of Algol —which Goodricke discovered without the use of a telescope and working in the English climate— but also determined the period of this light variation. He even gave two theories as to the cause and one of these —the periodic passing between us and the star of a "dark body"— is essentially correct. Indeed, the other possibility —the existence of highly variable star spot areas is now used to explain certain details of the changes in a few other close systems. Goodricke even searched the earlier literature and found that Montanari and Miraldi had much earlier discovered the variability but not the periodicity, and he gave them credit. Since 1783, the study of eclipsing binaries has been of ever increasing volume and no attempt will be made to give a detailed picture. I could at least mention that the eclipse hypothesis was firmly established when changes of radial velocities in eclipsing systems were determined.

There are several reasons for observing eclipsing binaries. If observations covering the entire light curve have been made, then methods exist of deriving from the light changes the relative sizes of the two stars and the inclination of the orbital plane to the line of sight. With these known, the study of the velocity curve can yield sizes, masses and densities —i.e., the fundamental data concerning the stars themselves and matters not directly obtainable from single stars. These determinations require observations of the entire light curve and in many cases re-observation to exhibit whether or not changes occur which are not necessarily connected with the eclipse effects. Account must be taken of the distortions of each star from a sphere caused by the gravitational effects of the other, the brightening (especially in the colder component) of the hemisphere facing the companion and frequently various other effects not completely understood. The exact "solution" can be a difficult and time consuming chore and is probably best left to the experts in the field. Further, in many cases, the wave-length region of observation must be carefully specified and this means standard stars, color differences and other difficulties.

However, there is one extremely valuable contribution which can be made by the amateur which is free from color differences and which is a contribution which will grow in value as time goes by. That is to say, all future studies will have to relate back to this and thus the present observations will be of increasing value the more the system is observed in the future. This contribution is the determination and publication of the times of minimum light of eclipsing binary systems.

The amateur astronomer wishing to observe these will probably do well to establish contact with one of the various organizations for the study of variable stars. These exist in many nations but their membership lists cross international boundaries. Many but not all of these have been described in the Journal of the AAVSO 15, No. 2, Pg. 141-161-75th Anniversary Edition. A few of these confine their work almost entirely to minima of eclipsing systems. (An excellent example might be noted by looking through the BRNO contributions).

One such highly productive organizations is the BBSAG (Bedeckungsveranderlichen Beobachter der Schweiezerischen Astronomischen Gesellschaft) whose present headquarters are c/o K. Locher, Rebrain 39, 8624 Grut, Switzerland. The membership consists of observers from a number of different countries. A major contribution from this association lies in the publication of the observed times of minima in a publication which receives world wide distribution in the astronomical community. The most precise observations are of little value if they remain only in the observing notes of the astronomer himself. Also, while I know that I do not have to tell this audience how to observe stellar brightness, the presentation of them in a form to be of maximum value requires the reduction of the observed times to Julian days and fractions of a day and correction of these heliocentric times so that the observation not be vitiated by the movement of the earth about the sun. These corrections must be applied before a meaningful period study can be made. For any astronomer not familiar with means of determining these, contact with an established group of variable star observers (such as the BBSAG) or with a major professional observatory will probably answer all questions. The publication Rocznik Astronomiczny of the Astronomical Observatory of the Jagiellonian University in Cracow is also of great value in selecting an observing program. Astronomers with photoelectric photometers might wish to get in touch with the IAPPP - International Amateur and Professional Photoelectric Photometry.

Perhaps now we should consider just why it is important to observe these times of light minima. Just what will we learn that will justify the long hours at the telescope? There are various answers.

One is simply the determination of better light elements. (The light elements are simply the time of one well determined light minimum-the epoch- and the period or length of time from one minimum to the next. These permit computation of future minima or phases at any given time.) This is of particular importance to spectroscopists or to photometric observers planning a program.

However, frequently it is not possible to fit all the observed minima with any given set of light elements. This indicates that there have been changes in the period and it is of interest to note very briefly what can be learned from these.

On a few rare occasions, the close "double" star is a member of a triple star system with a companion well removed from the binary. The changes of period are caused by the motion of the close binary around the center of mass of the system. This can give information especially as to the masses of the stars involved which is of interest in studies of triple star systems. In all cases, the variation of period as determined by the (O-C)'s —the difference between the observed and computed minima— when plotted against time show a sinusoidal fluctuation with those of secondary minimum following the same pattern as primary.

A case of more importance from the point of view of astrophysics is when the two components are moving in elliptical orbits about the common mass center. Without going into details it can be mentioned that in this case the line of apsides joining the stars will rotate and that the rate of this rotation will depend, among other things, on the "model" in which the star is built or, more specifically, on the degree of central condensation. This relation will cause a periodic change of period as observed from the earth; the (O-C)'s of secondary will show a similar periodic change but one exactly out of phase with the primary. Such systems can be detected most easily by the fact that the secondary will not occure half way between primaries but will be displaced. (A more complete discussion can be found in Photoelectric Astronomy for Amateurs, Ed. F.B. Wood, published by the Macmillan Company in 1963, and unfortunately now not always easily available). However, it takes many epochs of observation to establish whether or not the period variation is indeed itself periodic and not the result of one or two apparently sudden changes. One clue is that the (O-C)'s of the secondary minimum should also vary periodically but exactly out of phase with those of the primary.

The most common changes are apparently abrupt lengthening or shortening of the period which effects both primary and secondary alike. Again, the conventional treatment is the plot of the (O-C) curve against epoch. Also continuous observation is important. Nearly forty years ago it was pointed out (F.B. Wood, Ap. J.112, 196, 1950) that in almost all of these cases at least one component of the system was near the zero-velocity surface outside of which the star is no longer is in complete gravitaitional control. (For a more complete discussion of this see Interacting Binary Stars by J. Sahade and F.B. Wood, Pergamon Press 1978). Stars well within these limits generally show no such sudden period changes. The suggestion first made was that these changes could be caused by sudden, violent mass loss to the system; today, the idea of mass transfer from one star to the other is widely discussed. However the business of the observer is to make and to publish the times of minima. Since the period changes are usually rather small, observation over many years are usually required to establish the reality of the variation and even more to establish its nature. Thus the observer of today is indeed laying the basis for all future work.

The planning of an observing program is extremely important. The size of his telescope and the sensitivity of his detector will set magnitude limits. The latitude of the observer will set limits of declination. Stars with very long periods will present difficulties; observation on two or more successive nights may be necessary to obtain one well determined time of minimum. Yet this very difficulty means that the system is less frequently covered and therefore the observation of greater importance. Consultation with the organized observing groups, (AAVSO, BAV, BBSAG, BRNO, IAPPP, and others) will be of great help. A few copies of the fifth edition of "A Finding List for Observers of Interacting Binary Stars" (Publ. of the Dept. of Astronomy, University of Florida, Vol. I) are still available and will be supplied on request while they last. This gives also suggestions as to comparison stars because observations are always given as differences of brightness between the variable and one or more stars of constant light. For almost all systems at least rough light elements exist and these can be used to compute future minima so that the observer will not spend a lot of time observing phases of constant light. In addition to the scientific value, there is something exciting in seeing the star actually change in brightness with time.

In conclusion, the above recommendations are in no way intended to belittle the many other fields in which amateurs can make useful contributions to astronomy, but merely to call attention to the excitement and the permanent value of determination and publication of the time of minimum light of eclipsing binary systems.

All good wishes for skies of high photometric quality.

Amateur Astronomy in Poland: Past and Present

Katarzyna Turaj
Wyzsza Szkola Pedagogiozna, ul. Podchorazych nr. 2, PL-30084 Krakow, Poland

Introduction

Amateur astronomy in Poland has its own history dating from the 17th century to the present day. Before the 17th century, the term "amateur astronomer" loses its meaning. Otherwise we might pose the rather paradoxical question: "Was Copernicus an amateur?" and probably have to give the answer: "Yes, he was an amateur, being first a priest, a physician and a lawyer." Let us leave him in peace and instead turn to more undoubted amateurs. The history can be divided into two general periods: before and after the creation of the Polish Amateur Astronomical Society, PAAS (Polskie Towarzystwo Milosników Astronomii – PTMA). Here we present 16 Polish amateur astronomers who contributed to astronomy from the 17th to the 20th centuries, except Jan Heweliusz – the greatest – who is discussed elsewhere (1). All are selected from a much larger group, the selection being made in accordance with the rules described in the very useful and practical "Criteria for identifying an astronomer as an amateur", formulated by Tom Williams a few years ago and presented here (2). There is also a short history and current information about the PAAS. Finally, we summarize successes and failures of amateur astronomy in Poland and put some general questions about its future.

Great Individuals from Urania's Host

Let us begin with a lady. Maria Kunicka (1620–1664) is considered to be the first Polish woman astronomer. She was born at Swidnica and well-known for her erudition and beauty. Her most important work was *Urania propitia sive tabulae astronomicae* (Byczyn, 1654, by Johann Seyffertus the printer). There are probably some individual volumes in the oldest European libraries, but alas not in Poland.

Stanisław Rola-Lubieniecki (1623–1675), was an Arian minister (bishop), interested in explaining comet appearances in a scientific way. He wrote the 3-volume work *Theatrum Cometicum* (Amsterdam, 1667) about comets from "The Flood to 1665".

Józef Aleksander Jablonowski (1712–1777), Voivode of Nowgród and a prince. He wrote a historical and astronomical work to advance Copernicus' theory *De astronomiae ortu atque progressu et de telluris motu* (Rome, 1763). In 1774 he founded the scientific society called "Societas Jablonoviana".

Jerzy Kunowski (1786–1846), a lawyer born in Silesia, but it is not known whether of Polish or German origin. He studied the surfaces of the Moon and the planets, and also calculated the rotational period of Mars. There is a crater on the Moon named Kunovsky after him (400 km SE of Copernicus).

Hercules Dembowski (1812–1881), a sailor born in Milano as son of a Polish emigrant and an Italian aristocrat. He founded an observatory on the side of Mount Vesuvius and observed double stars there; estimating their positions was his main interest (3). His findings were published in *Astronomische Nachrichten*. Dembowski was awarded the Copley Medal of the Royal Society of London (1878) and the planetoid 349 was called Dembowska from his name. One can also find a crater Dembovsky on the Moon.

Jan Walery Jedrzejewicz (1835–1887). A physician, who appears to be the greatest Polish amateur astronomer of the last two centuries. He founded an observatory at Plonsk, where he made most of his observations of double stars. He measured positions of 16 comets and made systematic observations of sunspots and occultations of stars by the Moon. Besides publishing papers in *Astronomische Nachrichten*, he wrote a handbook of astronomy called *Kosmografia* (Warszawa, 1886).

Zygmunt Laskowski (1841–1928), a professor of medicine. His most important contribution was the discovery of Nova Aquilae 1918.

Adam Ostoja-Ostaszewski (1860–1934). This strange person could be called the "enfant terrible" of Polish amateur astronomy, although he was also called "the Leonardo of Wzdów" because of his aviation inventions. After building an observatory he published the mad, but fascinating work, *Le vrai système du monde*, putting in it his fictitious cosmology "proving" that the Sun is situated inside the Earth, and that one only sees its image in the mirror-like sky.

Feliks Przypkowski (1872–1951). A physician interested in time-keeping and clocks. He gathered a large collection of sundials of many types and ages. The collection forms the unique Museum of Clocks in Jedrzejow.

Antoni Wilk (1878–1939), a teacher of physics who became a professional astronomer and a renowned comet hunter. He discovered the following comets: 1925k, 1929d, 1930c, and 1937c. He was awarded four medals by the Astronomical Society of the Pacific.

Tadeusz Rakowiecki (1878–1965), a practising physician who was self-taught in astronomy. He used to say that he had not become a professional astronomer because of the need to serve people more directly. He turned down an offer to stay at the university as a professor of astronomer. He wrote 24 papers on the calculation of orbits, double stars and eclipse theory, and also published the astronomy handbook *Ways of planets and comets* (Warszawa, 1928).

Antoni Rybarski (1889–1972) was an engineer interested in amateur telescopes. He built a 350-mm Newtonian and a 150-mm telescope combined with a camera for photography. He also invented a telescope with a metal mirror (PL patent No. 42426-1959).

Feliks Rapf (1891–1972). A teacher of physics interested in education and the popularization of astronomy. He organized many courses in astronomy and published papers and notes in *Urania* and the *Physics at School* magazines.

Jerzy Pokrzywnicki (1882–1974). A lawyer, who engaged in scientific research into meteorites. He described many meteorites in his 60 papers.

Władysław Lis (1911–1980), a janitor at the Lubomir Observatory (in the Beskidy Mountains). He discovered a comet known as Kaho-Kozik-Lis (1936 III) and was awarded the J.A. Donhoe Medal of the Astronomical Society of the Pacific.

Edward Szeligiewicz (1924–1959), an architect who observed meteors, solar and lunar eclipses, and variable stars. He left two interesting manuscripts: *Atlas of Eclipsing Binaries* and *Small Atlas of the Moon*.

The Polish Amateur Astronomical Society

Polish amateur astronomy began a new phase in 1919 when several keen youngsters from secondary schools started the small "Amateur Astronomical Circle" with about 20 members. They published four issues of a magazine *Uranja*, printed by lithography. In 1921, the Circle was legalized and changed its name to "Polish Society of Friends of Astronomy", being officially founded by adults. When it began it had about 50 members. That number increased to about 3500 today. Before World War II, members of PFSA organized many lectures and sky demonstrations for the general public. The quarterly *Urania* published astronomical information and data, and popular articles under the title of "Mysteries of the Universe". During the War, in an occupied country, the PSFA did not exist as a formal organization. It was a time when even professional astronomers worked as amateurs. As for young amateurs, two groups continued their activity: in Dobra Woda (editing reports and the list of objects observed), and in Sporysz (printing the clandestine magazine *Mathematics and Astronomy Monthly*, and organizing an expedition to watch a lunar eclipse in 1942. This seems to be an interesting case, proving that even astronomy can be underground. (4)

Amateur astronomers reactivated their Society in 1948, but had started even earlier by organizing some popular lectures on Polish radio in 1945, and by publishing the first post-war issue of *Urania* in 1946. Thenceforward the association was named the Polish Amateur Astronomical Society (PAAS).

The Activities of the PAAS

The society acts in the form of clubs in various towns, these are equipped with instruments according to their own resources – ranging from just binoculars to even a small observatory. The clubs are supervised by Headquarters, which undertakes general work, such as: training observers; coordination of amateur research; editorial activities and the distribution of astronomical publications. The clubs arrange lectures, star parties, film or discussion evenings, and keep small astronomical libraries. Camps and courses for beginners or advanced observers are arranged by Headquarters, paying particular attention to the education of the young. *Urania* is published regularly, while the *Scientific Supplement* to *Urania* – the astronomical reports – other leaflets, charts, and calendars are published from time to time. Observers work in six sections: Variable Stars; History; Meteors and Meteorites; Positions and Occultation Timing; Solar; Sky Patrol. The most active nowadays seems to be the Positions and Occultation Timing Section, which is a member of I.O.T.A. This is significant, because the Section of the PAAS was the organizer of the European Symposium on Occultation Projects in 1986. The Solar Section, called "The Sunflowers" is also active, but has recently split away from the PAAS and called itself the Solar Observers' Society.

The Sections cooperate directly with professional astronomers, partly because of the need for some help and advice, and partly because of the need to publish findings in scientific journals. It should be stressed that the PAAS, although an amateur society, has never closed its door to professional astronomers. On the contrary, many professional astronomers, especially those fond of popularization, are active and devoted members of PAAS, and devote a lot of work to Headquarters and the Scientific Council.

The Present, and Some Questions About the Future

The PAAS now consists of 30 clubs. Several are still very active, but there are others that are slowly dying, because of the small number of active members. Does this mean that astronomy is losing its attractiveness? Many letters from youngsters that arrive every day at the PAAS Headquarters show that it does not. Headquarters do their best to answer questions either by post or by organizing courses. But many young people find the courses disappointing. Lectures cannot compete with amazing scientific programmes often shown on television. Some old forms of amateur activity should be changed to modern ones. On the other hand, the development of amateur astronomy today depends greatly on technical equipment – if we assume that making observations is the main way for amateurs to make a real contribution to astronomy. Overall achievements are hardly limited by the level of technology, and the gap between highly-, and poorly-developed countries is growing as time goes by.

The PAAS continues the tradition of great amateur astronomers in a country that is at a low level of technology. But there is also the other aim of amateur astronomy, which appears somewhat less attractive. There the aim is to be a link between high-level professional science and ordinary life, transmitting information from the very exclusive group of scientists to the general public. That purpose – with a little help from the modern mass-media – could probably be developed in any country in the world.

1. Iwaniszewska, C., "Johannes Hevelius: Polish seventeenth-century brewer and astronomer", contributed paper presented at this Colloquium
2. Williams, T., "Criteria for identifying an astronomer as an amateur", contributed paper presented at this Colloquium
3. P. Couteau, "The history of amateurs' contribution to the study of double stars", invited paper presented at this Colloquium
4. [*See also*, Walter, K., "Astronomy in Poland during the Second World War", *J. Brit. astron. Assoc.*, **97** (5), 270, (1987) – Eds.]

Giovanni Battista Lacchini: An AmateurAstronomer from Italy

Giancarlo Favero and *Sandro Baroni*
Largo E. Fermi 5, I-50125 Firenze, Italy

Twenty years ago, on the 6th January 1967, the astronomer Giovanni Battista Lacchini died in Faenza, a town near Bologna where he was born 82 years before, on the 20th May 1884. But, as we shall see, he spent most of his life very far from Faenza, moving his residence all around the Italian peninsula and also over other countries.

After completing his studies, the 22-year-old Lacchini was appointed bookkeeper at the Mixed Court of Cairo (Egypt). Two years later, called as a postal office employee, he came back to Italy. During that period, reading the Camille Flammarion's production, Lacchini learned to read the poetry of the sky. Since then, he devoted most of his spare time to the study of astronomy and to the observation of any astronomical event falling under his eyes. Soon, he found the most congenial field of application in the study of the variable stars.

Lacchini was one of the earliest members of the American Association on Variable Stars Observers, to which he contributed more than 53,000 magnitude determinations between 1911 and 1963. His contribution was so outstanding that AAVSO lent him a 8 inch reflector for his private observatory. In fact, Lacchini made most of his observations by naked eye or using small instruments (e.g. binoculars or refractors). He always carried some of them with him during his frequent journeys around Italy. Moreover, he often performed a timely observation through the open window of a moving train. He became so skilled in evaluating the magnitude of a star that he could estimate the figure to nearest decimal using only one comparison star !

Some variable stars, such as T Orionis and Z Camelopardalis, were observed for a long time only by Lacchini : for that and other valuable contributions his work received international recognition in 1922, when he became a member of the Variable Stars Commission of the International Astronomical Union. Contrary to the rule *"nemo propheta in patria"* the outstanding contribution of Lacchini to the progress of astronomy was recognized in Italy. In 1928 a special law was promulgated which appointed him to the staff of the Italian astronomical Observatories. Previously he had been to Catania, where the clear sky allowed him to carry out a great number of solar and variable stars observations. From 1930 he was at Pino Torinese, the Observatory of Torino University, where he discovered a new asteroid, 1930 AB, and carried out many observations of telescopic meteors. Finally, in 1933, he moved to Trieste Observatory where he observed novae, comets, double stars and lunar occultations.

After his retirement, in 1952, Lacchini returned to Faenza and became a valuable co-worker of Guido Horn d'Arturo, the Director of the Bologna and Lojano Observatories. In this period he discovered tens of variable stars, analyzing the plates taken by Horn d'Arturo and himself at the moving focus of the fixed multi-mirror reflector of 1.8 m diameter, which antedated by thirty years the present interest in giant multi-mirror instruments.

During his entire career Lacchini enthusiastically contributed to the popularization of Astronomy, following his ideal, Flammarion. He held many conferences and wrote hundreds of papers, published mostly in the journal "Coelum", edited in Bologna by Horn d'Arturo. Virtually all Italian amateur astronomers possess his most popular book, the "Atlante Celeste" covering all the naked eye stars to magnitude 6.5 and which had four editions : 1948, 1954, 1960 and 1969. Another important publication was the "Atlante Celeste spettroscopico", reporting the spectral classification for all the naked-eye stars.

The example of Lacchini stimulated a number of Italian amateurs, who devoted their spare time to the study of variable stars. Among these we find Luigi Jacchia, Giuseppe Loreta, Giovanni Bernasconi (better known for his work on comets), Giuliano Romano (discoverer of hundreds of variables and three supernovae), Luigi Baldinelli, Marino Perissinotto and Italo Dalmeri. Let us hope that these examples will be followed by many amateurs from later generations.

The Instruments of H.E. Dall

E.J. Hysom
8 East Drive, Caldecote, Cambridge CB3 7NZ, U.K.

Author's note: Due to lack of space, biographical details have been largely omitted – these can be found elsewhere (1,2,3). It should be noted that Horace Dall spent most of his working life as an experimental physicist with George Kent Ltd, fluid-flow meter manufacturers; he was well-known for his skill in developing prototype instrumentation.

Optics

In 1926 Horace Dall commenced optical work, with interests covering both microscopes and telescopes. Soon (1928), he was making solid eyepieces of the Tolles type. Eventually he developed skills in making lenses of the smallest size to be incorporated into microscope objectives of the highest numerical aperture; this culminated in an objective with a N.A. of 1.92 (a record). The lenses were jewelled elements worked with diamond dust. During World War 2 he repaired all the microscope lenses damaged in the U.K. that had originated with the German firm of Leitz.

Horace Dall had an exceptionally inventive mind – and was so active in developing his ideas that he had insufficient time to formally write them up, hence much of his work remains unpublished. He did, however, keep extensive notebooks of his researches into optics. These, along with many prototypes, are now in the custody of the Science Museum, London. Combining his interests in travel, optics, and astronomy, Dall developed a number of portable, ultra-lightweight telescopes. He discovered that by incorporating what is in effect a long working distance, low-power microscope with a Cassegrain telescope he had an instrument with several advantages. The image was erect, the secondary small, and by use of a suitably placed internal stop, sky-flooding was eliminated without the use of shade tubes. Always looking for improvements in optical performance he hit on the idea of modifying the classical Cassegrain telescope by emplpying a prolate ellipsoidal primary mirror with a spherical secondary: independently discovered by Kirkham in the U.S.A. the type is now universally known as the Dall-Kirkham.

As an experimental physicist Dall was a great advocate of null tests – his null test for paraboloids must have resulted in tens of thousands of higher quality mirrors being made than would otherwise have been the case – this test was later extended to cover prolate ellipsoid and hyperbolic concave mirrors, making it immensely valuable to the optical worker. Later he developed several null tests for hyperbolic Cassegrain secondaries.

Because of his deep knowledge of optics – most of it self-taught – Dall was often used as a consultant by professional telescope makers; additionally he helped many thousands of amateurs with advice or with testing, repairing or re-figuring their optics. Finally mention must be made of some of his other achievements – his early use (1939) of thin-edged primary mirrors; his micro-writing, of letters one wavelength of light high using a single diamond point on glass; of his development of the world's most accurate mechanical spherometer – it could detect 1/20th wavelength variations directly. He also developed an H-α telescope – the Promscope – that has found favour with solar observers. Now that the world of instrumentation is moving in different directions it is unlikely that the world will see his like again.

References

1. *Sky & Telescope*, **72**, (Oct. 1986), 410–411
2. *J. Brit. astron. Assoc.*, **97** (2), 76–80, (1987)
3. *Telescope Making*, **28**, 4–9

Roberts of Lovedale and Eclipsing Binary Stars

Thomas R. Williams
3621 Wakeforest, Houston, TX 77098-5509, U.S.A.

Alexander William Roberts, a great humanitarian and teacher in South Africa, was also a luminary in the astronomical world. This paper discusses his work on variable stars and touches briefly on his career in other areas. Born and educated in Scotland, Roberts migrated to South Africa in 1883, at age 25, to teach at the Native College at Lovedale. He studied mathematical astronomy as a recreational pursuit, but became an active observer in 1889. After two years of general observing, be began a systematic survey using binoculars and an old one-inch theodolite. He carefully plotted all visible stars in selected areas and ranked them in order of their apparent brightness, doing so repeatedly on six evenings for each of the selected areas. His composite sketch became a reference chart as he searched for changes in these fields. With this technique, Roberts discovered more than 20 variable stars.

Early in his observing career, Roberts emphasized the importance of technique in visual photometry, a topic which he pursued in several papers on position-angle and field-orientation effects. In practice he removed the effect of the relative field position of the variable star and the comparison star by using one normal eyepiece and another eyepiece that reversed the field. He then average the two results on each observation. Eventually, he averaged four to six estimates made from various position angles with a special prismatic telescope designed for precise visual photometry. His analysis of light curves, developed from these careful observations, formed the basis for his most significant contributions in eclipsing binary astronomy.

Roberts' contributions to eclipsing binary astronomy are illustrated in the controversy on V Puppis. This star's variability, first suspected by the English amateur astronomer Arthur Stanley Williams in 1886, was confirmed by E.C. Pickering in 1895 with a period of 3.445 days based on spectroscopic observations. However, Roberts published his period of 1.4544757 days in the April 1901 *Astronomical Journal*. Roberts apparently came under some pressure from other astronomers to explain the period discrepancy since the spectroscopic period was generally accepted as the final authority. However, he resolutely stood his ground, stating "I have spoken my last word on the subject". The question remained unresolved until 1905 when Pickering stated in a letter to Roberts that "our observations of the spectrum of V Puppis agree with the period you found for that star." The Third General Catalogue of Variable Stars (1970) gives 1.4544877 days as the period for V Puppis, in good agreement with Roberts' original value. The resolution of the V Puppis period was important in removing a cloud over other of Roberts' conclusions based on the same observational data, especially his 1901 *Astrophysical Journal* observation that the V Puppis stars were nearly in contact and were severely distorted into ellipsoidal shapes by gravitational tidal effects. His conclusion provided crucial observational proof of theories which had been proposed separately by Poincaré and Darwin on tidal distortions of the shapes of stars and other astronomical bodies.

Roberts made many other contributions regarding the shapes and orbits of binary stars. In an 1899 *Astrophysical Journal* article, he calculated the upper limits of the density of the individual stars that are components of four southern eclipsing binaries in which the stars are very extended and tenuous objects. He showed that a circular orbit establishes an upper limit on the average density of the stars. His upper bound densities are similar to those proposed by Henry Norris Russell in the same *Astrophysical Journal*. The value of Roberts' work on binary stars has been acknowledged by his contemporaries as well as modern binary-star astronomers including W.W. Campbell, Zdenek Kopal and V.P. Tsesevich.

While gaining recognition as an astronomer, Roberts' reputation was growing as a teacher and humanitarian. In his main profession as a teacher at Lovedale College for Natives, he was personally responsible for training over 4000 native teachers. In 1920, Jan Smuts, Prime Minister and Minister for Native Affairs, chose Roberts to represent the natives as a white Senator in the all-white Parliament, and appointed him to a three-man Native Commission and several special investigative commissions. Poor health and political considerations forced Roberts to retire from these positions in 1935. Roberts was recognized as the foremost authority on native questions in South Africa, and would deserve a secure place in the history of South Africa based solely on his humanitarian efforts. Thus his significant contributions to astronomy are all the more interesting. In the words of Jan Smuts: "After Roberts let no amateur despair, and let each cultivate his scientific hobby to the utmost limit of his powers and opportunities. Science ... owes much to the amateur. Roberts was truly a prince among our scientific amateurs."

Reference

Williams, T.R.: *Alexander Roberts: distinguished South African amateur astronomer and humanitarian*, Abstract 01.07, Bull. Am. astron. Soc., **17** (4), 1985

An full bibliography regarding Roberts may be obtained from the author.

Seventeenth-Century Solar Observations: Fundamental Results with Amateur Instruments

E. Ribes-Nesme, J.C. Ribes
Observatoire de Paris, Section d'Astrophysique,
5, place Jules-Janssen, F-92195 Meudon Principal Cedex, France

Abstract

The equipment used by professional astronomers in past centuries was no better than that used by modern amateurs. Systematic observation of the Sun began in the 17th century and continuity of observations is very important for understanding the solar cycle.

Results obtained from historical observations were presented, in particular covering the Maunder Minimum.

Continued systematic observation is important for the understanding of long-term solar-terrestrial relations.

References

Ribes, E., Ribes, J.C., Barthelot, R.: "Solar diameter and solar rotation during the Maunder minimum" in *Proceedings of Helio-asteroseismology, Aarhus meeting*, Christensen & Dalsgaard (eds.), 1986

Ribes, E., Ribes, J.C., Barthelot, R.: Nature, **326**, (No.6108), 52–5, 1987

Ribes, E., Ribes, J.C., Barthelot, R.: "The solar envelope has expanded during the Maunder minimum: by how much?", in *Proceedings of Vancouver meeting, I.G.G.A., August 1987*, in press

Ribes, E., Ribes, J.C., Barthelot, R.: "On the accuracy of Picard and La Hire's observations of the Sun and Jupiter during the seventeenth century", *Nature*, in press, 1988

Sailors and Astronomy

Antoine Demerliac
48 rue dun Moulin-à-Poudre, F-29200 Brest, France

Sailors are neither professional nor amateur astronomers, but people who have to observe the sky in the course of following their professional calling. As soon as sailors ventured far from the coast they required some guide and even in antiquity (as shown by mentions in Homer's *Odyssey* and, considerably later, by Aratus) they used Ursa Minor and Ursa Major. With journeys farther afield, they gained some insight into latitude from the changes of the altitude of the circumpolar constellations above the horizon.

Although the very last Viking voyagers knew of the magnetic compass, earlier ones still relied on the Sun and the stars. Describing the methods used, Harald Akerlund says "One saga mentions a man, Oddi Heldagon, who was known as "Oddi the Star" and who served as long-distance pilot for an Icelandic magnate towards the end of the 900s. He left notes that included a complete table of the changes in declination of the Sun throughout the year, expressed as the height of the Sun on the meridian in semi-diameters. There is also a small table of azimuths giving the direction at different times of the year, of dawn twilight, defined as being a faint band of light on the horizon, visible before sunrise. We know nothing of the instrument used to measure the altitude of the Sun."

Such methods fell into disuse with the discovery of the compass, after which positions were estimated from the known point of departure and the distance covered. Such methods sufficed for navigation in the Mediterranean and around the coast of Europe, but when Henry the Navigator instigated longer voyages of exploration around Africa, estimation of position on the high seas and at low latitudes required new techniques. The solar declinations given in the Alfonsine Tables, prepared between 1248 and 1252, were insufficiently accurate to be used with determination of the altitude of the Sun at local noon for position-finding. Abraham Zacuto of Salamanca published his more accurate *Almanach Perpetuum Celestiam Radix* in 1473 and Regiomontanus his *Ephemerides Astronomicae* in 1475. The Portuguese appear to have used accurate tables after 1449, but these remained a "trade secret" until their publication in Lisbon in 1509.

Although latitude could thus be determined, longitude either required a suitable chronometer or the knowledge of the precise times, at a standard meridian, of phenomena such as the eclipses of Jupiter's satellites, or else tables of lunar distances. The observatories of Paris and Greenwich were set up specically to provide accurate astronomical tables. The work of such observatories was wide-ranging and so cannot be described in any detail. However, the career of Admiral Mouchez is interesting, in that after having served at sea he became Director of the Paris Observatory.

In the middle of the 19th century, Lieutenant Ernest Mouchez found errors in both charts and in the astronomical methods and equipment used by navigators. In particular, a sextant used with a mercury artifical horizon did not give sufficient accuracy to check chronometers. Mouchez felt that sailors required the sort of techniques and instrumentation available at a land observatory. At his own expense he had a portable transit telescope built by Brünner, and this was tested on the voyage around the world that he made on the "La Capricieuse" between 1850 and 1854. On his return, his report about observations made with the instrument, its drawbacks and possible improvements, was sent on to Le Verrier at Paris Observatory, who then invited Mouchez to develop the instrument at the observatory. This perfected design of portable transit telescope remained in use until the development of the prismatic astrolabe.

Mouchez also designed an altazimuth telescope, which again he had built at his own expense by Brünner. Both instruments were used by Mouchez to determine the longitude of the observatory at Rio de Janeiro in Brazil. His value disagreed by at least 21 seconds – equivalent to a distance of more than 8 km – from the value determined by the director, Emmanuel Liais, who was seconded from Paris Observatory. A commission of enquiry essentially agreed with the value found by Mouchez, who was appointed to the Bureau des Longitudes in 1873.

Naval officers took part in the expeditions to observe the transit of Venus in 1874. Bouquet de la Grye led that on Campbell Island and Mouchez the one on Saint Paul. On his return to France, Mouchez suggested that the scientific equipment that had been brought back should be used for instructing naval officers and explorers. The Navy, particularly its Chart Office, the Institut, the Académie des Sciences and the Bureau des Longitudes were all interested and the City of Paris donated land, where the Montouris Observatory was established. This was directed by its founder until his death, and then passed to the central hydrographic service and later the Bureau des Longitudes, until the Navy ceased to send its officers there for training.

Mouchez was elected to the astronomical section of the Académie des Sciences in 1875. In 1878 he became Director of the Paris Observatory after Le Verrier's death and after Henri Faye, who was proposed by the Académie des Sciences, had been rejected for unknown reasons. He was seconded by the Navy to take up this post and appointed Rear-Admiral at the same time.

The second transit of Venus, in 1882, saw Bouquet de la Grye and Hatt leading expeditions to Puebla in Mexico and to Chubut in Patagonia. The former was the third President of the S.A.F., after Flammarion and Hervé Faye.

Two other hydrographic surveyors, Caspari and Fichot, became Presidents of the S.A.F., and both are better known for their theoretical work than for their observations. Caspari, who wrote a manual on astronomy, worked on improving time-keeping, and published a large number of papers about chronometers. He was President of the International Commission on Time-keeping that met at the time of the Paris Exhibition in 1900. Finally, Fichot, apart from numerous scientific papers, edited one of the volumes of Henri Pioncaré's manual on celestial mechanics.

Bibliography

Hogben, L., *Science and the Citizen*, London, 1938 (3rd edn 1951)

Lallemand, F., ed., *Journal de bord de Pytheas [Pytheas's log]*, Editions de Paris, 1956

Olagué, I., ed., *Journal de bord de Jean de la Cosa, second de Christophe Colomb [The log of Jean de la Cosa, Christopher Columbus' second-in-command]*, Editions de Paris, 1957

Charcot, J.-B., *Christophe Colomb*, Flammarion, 1934

Banbury, P., "A la recherche des premiers marins", Editions France-Empire, Paris, 1977

Phillips-Birt, D., "L'art de naviguer, de la préhistoire à nos jours", Editions Maritimes et d'Outre-Mer, Paris, 1975

Taylor, E.G.R. & Richey, M.W., *The Geometrical Seaman*, London, 1962 Bouquet de la Grye, "Mémoires relatif au passage de Venus sur le Soleil en 1874 et 1882", (held by the library of the Institut)

Mouchez, R., *Amiral Mouchez, Marin, Astronome et Soldat*, 1821–1882, Editions Cujas, Paris, 1970

The Map of the Heavens at the Chateau de Saint-Jean-de-Chepy

Joël Barbier
AEROSPATIALE, Sciences et Archéologie,
100 Bd. du Midi, B.P. 99, F-06322 Cannes-La Boca, France

Abstract

In the Dauphiné, near Tullins, there is a fortified manor-house known as the Chateau de Saint-Jean-de-Chépy, built in the 13th century, that has in one of its towers an Italian Renaissance map of the heavens painted on the ceiling. The south tower could have been designed for astronomy, with the octagonal-roofed, second floor with its seven windows being used for observation. The first and ground floors have hemispherical vaults, and painted on the ceiling of the first floor there is a map of the heavens that is unique in France. It is painted on the ceiling, with the southern constellations extending down onto the cylindrical wall. It is similar to Italian Renaissance maps, like those in the Old Sacristy at San Lorenzo in Florence (painted by Brunelleschi in 1429), in the Villa Farnesine at Rome (Peruzzi, 1511), and in the Villa Farnese at Caprarola (artist unknown, 1575).

An owner of Saint-Jean-de-Chépy was Maurice Bressieu (1546–1617), a distinguished mathematician, holding the mathematical chair at the Collège de France between 1575 and 1586. He wrote *Metrices Astronomicae*, which was "highly regarded by mathematicians" and became Speaker for the Kings of France at the Vatican. Pope Sixtus-Quintus made him steward of the Vatican Library. It therefore seems very likely that the map could have been painted for him.

Amateur Astronomy in France, 1789–1830: Two Examples

Simone Dumont[1] and *Solange Grillet*[2]
[1]Laboratoire d'Astrophysique Théorique, IAP
98 bis Bd. Arago, F-75014 PARIS, France
[2]Observatoire de Paris, 61 av. de l'Observatoire,
F-75014 Paris, France

Abstract

Historical events affected astronomy in France at this period. Large instruments were also rare so amateurs were on equal terms with professionals. Two significant amateurs are described:
Honoré Flaugerges (1755–1830), justice of the peace at Viviers (Ardèche), observed from 1782 until his death, in particular discovering two comets and recording sunspots.

Anne-Jean P.C. Duc-Lachappelle (1765–1814), studied astronomy in Paris, and set up an observatory in Montauban in 1789, mainly observing planets and Mercury in particular. He had one pupil, Bernier, and set up a local society for arts and sciences in 1796.

Both of these astronomers had numerous contacts with professional astronomers throughout Europe.

Some French Amateurs of the 2nd Half of the 19th Century

Françoise Hourmat
Residence Concorde, 13 square St. Florentin, F-78150 Le Chesnay, France

At that period there was keen popular interest in astronomy, and in many Paris squares, astronomers with terrestrial refractors gave talks on astronomy for a small sum. Léon Joubert created a observatory for scientific research and popularisation, allowing anyone to learn about the universe and use good instruments. He made 120 instruments: refractors, reflectors, projectors, and photographic instruments.

Hermann Goldschmidt (1802–1866), born at Frankfurt am Main 17 June 1802, had poor health, became a painter and sought his fortune in Paris. He became an astronomer by accident after following a course of lectures at the Sorbonne given by Le Verrier. From a modest studio on the 6th floor of an old house in the heart of Paris, he discovered 14 minor planets between 1852 and 1861, the first being called Lutetia by Arago.

At Vaison (Vaucluse) Father Gaudibert (1823–1901) made observations of the Moon and various discoveries about rilles and craters that were published in *Mechanic, l'Astronomie* and the *Bulletin* of the S.A.F.

The doctor Edmond Lescarbault (1814–1894) lived at Orgères (Eure et Loire), 100 km west of Paris and built a small observatory on his house. On 26 March 1859 he saw a black dot against the limb of the Sun. Its passage across the Sun took 1 hr 17 mins 22 secs (siderial time). He thought that he had discovered the planet sought by Le Verrier to account for the perturbation of Mercury, but no one has ever seen the object again.

Amédée-Victor Guillemin was a journalist and scholar, who published twenty popular science books. He was up-to-date with the latest progress because he corresponded with research workers. His clear style and his well-illustrated book greatly helped to popularize astronomy.

Comet Halley in 1910 in Franche-Comté: or Collective Memory at Fault

A. Heurtel, M. Rougeot & J.P. Parisot
Association Astronomique de Franche-Comté, Parc de l'Observatoire, 34, av. de l'Observatoire, F-2500 Besançon, France

When old people are questioned they often say that they vividly recall the appearance of Comet Halley in 1910. When, in 1985, we started to analyze that previous apparition, we fully expected to find remarkable accounts in the newspapers. But instead of dramatic stories of its appearance on the night of 1910 May 18–19, we find nothing was observed until May 28, because the sky over a large part of France (even Paris), was cloudy. Even then, it was described in terms such as "This wretched comet has a very over-inflated reputation, it appears more like a nebula than the object that we were led to expect." So why then the popular excitment?

For three reasons: first, the arrival of the comet was repeatedly mentioned in the newspapers following its recovery in 1909 September, together with tales of what might be expected in May when it brushed past the Earth; second, there was the unexpected appearance, in January, of the comet known as the Johannesburg Comet, or the Great Daylight Comet of 1910, which was very prominent indeed. Thirdly, at about the same time as that comet was visible, there were disastrous floods all over France, as is shown by the commemorative plaques that can still be found in many different places. There was considerable speculation in the press about a possible connection between the comet and the floods and even Flammarion wondered if there was a hidden link. Speculations (and fears) became rife in May as Comet Halley steadily approached the Earth.

When Comet Halley proved to be so insignificant, it is hardly surprising that it became confused in people's minds with the much more dramatic Daylight Comet. The association of the latter with the disastrous floods, and the idea that there might have been some connection, was enough to ensure that the mistake was perpetuated. This is probably how legends come to be born.

Newspapers and journals of the period consulted: *Le Petit Comptois; L'Eclair Comtois; Le Pays de Montbéliard; La Dépêche Républicaine; La Brigade de Fer; Les Nouvelles Comptois; Le Journal de Belfort; La Frontière; La Franche-Comté à Paris; L'Astronomie*
A complete account is given in: *L'Astronomie*, 100, July/August 1986.

A Renowned, Yet Forgotten Astronomer: Marcel de Kerolyr

André Richardot
Observatoire de Sabarat, F-09350 Daumazan-sur-Arize, France

Marcel Bonnemain de Kerolyr was born in 1873 and became a talented professional violinist. He married a singer and acting as accompanist to his wife, toured Europe in the period before the 1914–18 War, even giving a recital before Tzar Nicholas II and the Tzarina. Between tours, they lived in a mansion on the Promenade des Anglais at Nice, and where they entertained a lot. It was in leaving their home, and under their very eyes, that the American dancer Isadora Duncan was the victim of a fatal accident in 1927.

But then catastrophe: his wife lost her voice, and was found to be suffering from chronic anaemia. No more tours, and no more fees. They managed to subsist by selling furniture, pictures and jewellery. De Kerolyr was profoundly depressed and even toyed with the idea of suicide.

He had already visited Nice Observatory, but his enquiring mind also brought him into contact with an amateur, who had retired to Antibes, Georges Raymond, who encouraged and advised him. He acquired some objective lenses and got in touch with a mechanic. He decided to devote the rest of his life to astronomy, particularly to astronomical photography.

They scraped together what money they had and bought a house outside Digne (Alpes de Haute Provence). The site was excellent (without, in those days, light-pollution) and most of the time there was almost perfect atmospheric transparency.

His observatory (illustrated in *L'Astronomie*, Oct. 1928), had a run-off roof and was constructed of wood and galvanized iron. It contained a home-made mounting, with a fine bronze worm-wheel and worm, but the rest of the mechanical parts (gear-train, transmission and a hand drive requiring one turn per second, was made from *Meccano*.

The instruments were: a camera with a portrait lens, by Derogy, diameter 130 mm, f/4.6; a camera with a Berthiot aerial lens, diameter 120 mm, f/10 (both equipped with 13 × 18 plate-holders); and a 250-mm, f/15 guide telescope.

I met him in 1928 and for two years, was at his home three times a week, at least! Mme de Kerolyr, who was then more or less disabled (she weighed 210 kg!), used to greet me with unfailing kindness and directed me either to the study or to her husband's observatory.

Probably through the intermediary of Mme Gabrielle C. Flammarion, there was some simple agreement between the Lumière laboratories and de Kerolyr, who obtained plates, paper and chemicals direct from Lumière.

The results can be seen in the photograph published in *L'Astronomie*, Oct. 1928.[1] This photograph of the Milky Way (exposure 2 hours) is crossed by the trail of an exceptionally bright fireball – estimated by Emile Touchet as being magnitude −12! The plates were slow: the America nebula is beginning to be visible, but the Pelican has not yet been recorded. A photograph of Messier 31 (*L'Astronomie*, May 1929 taken in May, 1929), with 7 hours of exposure over three nights, shows that the method of resuming the exposure and ensuring that the star images fall in exactly the same place had been perfected.

In 1930, he sold his house in Digne to my mother, and moved to Forcalquier, 50 km southwest of Digne, close to Saint-Michel, where the Observatoire de Haute-Provence was later to be built. Seeing the results from Digne, particularly those with the Berthiot objective, Ernest Esclanglon (Director of Paris Observatory), André Danjon (Director of Strasbourg and later Paris Observatories) and André Couder felt that they had just the man they needed to test sky conditions on the Forcalquier plateau and its surroundings. André Couder made the optics, and G. Prin the mechanical parts, for a 81-cm testing telescope. Initially, de Kerolyr worked with the instruments from Digne. For a photograph of nebulosity in Orion's belt (*L'Astronomie*, Sept. 1932) the exposure was 12 hours, in four sessions. The Opta and Super-Opta plates from Lumière were called "ultra-fast", but must have crawled along at about 10 to 15 ASA. Guiding was always by hand and included numerous interruptions.

The Couder-Prin telescope (*L'Astronomie*, Sept. 1933) was installed in its 8-m diameter dome, close to the de Kerolyrs' villa. Naturally I went there as often as possible, either by train or by bicycle. Many details of this telescope are given in Danjon and Couder's classic *Lunettes et Télescopes*. He then sold me his equatorial mounting and the Berthiot and Derogy cameras.

In May 1933, *L'Astronomie* published the first (IC 405) of several plates reproducing deep-sky photographs obtained by de Kerolyr with this near-perfect – and perfectly used – instrument. These plates were expensive to reproduce and, up to the last, published in 1938, were described as "having been paid for by X...". After his death, it was made known that the anonymous donor was Comte Aymard de la Baume Pluvinel.

The observatory was called the "Paris Observatory Astrophysical Out-station". Costs were met by Paris, and de Kerolyr worked there as "observer", with complete liberty of choice as regards subjects and methods. He was not an official employee. This was why, later, and at the instigation of Mme Flammarion, André Danjon and Dr Marc Badel, there were difficulties in getting him a small pension.

Over the months and years, photographs followed one another, and many were published in our society's journal. Some of these photographs were used by de Kerolyr to illustrate Jean Giono's *Le Poids du Ciel*, 1938. In 1937, the new Palais de la Découverte, built for the Exposition Universelle, showed a collection of them,

[1] Regrettably, few of de Kerolyr's plates can be located, and those available for reproduction do not include all the striking ones mentioned in this paper. We strongly recommend that readers should consult the relevant issues of *L'Astronomie*. – Eds.

crowned by an enlargement of Messier 31 that was 7 metres long, and full of wonderful detail.

The southern part of the America nebula was shown in one photograph (*L'Astronomie*, Nov. 1933). Several points deserve to be mentioned. First, the exposure-time was 20 hours. This brought out details hitherto unknown. The sky background, despite the long exposure, shows no fogging; where it appears slightly illuminated, there is actually faint nebulosity. In some parts the background is completely black. The "ultra-fast" plates of those days that are so very, very slow to us had the advantage of high resolution and high contrast. The faintest stars are about 20th magnitude, and their size on the original plate is about one-thirtieth of a millimetre. In *L'Astronomie* 1933, p. 516, de Kerolyr wrote "This shows the optical and mechanical perfection of the equipment, particularly if one considers the length of the exposure, and if one remembers that the plate-holder had to be removed some forty times to ensure that focus was exact (by Foucault testing)." I hardly know which to admire most: the perfection of the telescope, or the skill of the man who was using it!

The last photograph from our "observer", appeared in *L'Astronomie* for 1938 April. It was obtained by using a Zeiss objective, 130-mm diameter, collimated by Couder, stopped down to f/9, and mounted piggy-back on the telescope. It shows the very extended, faint nebulosity around the Pleiades, which with their immediate surroundings are intentionally overexposed, and completely burnt out. But faint nebulosity appears over an enormous area (the photograph covers 9° 16′ by 7° 14′; north is to the right). The reproduction was made from a composite of two images, one totalling 18 hours of exposure (1937 December 30–31 and 1938 January 1–2) and the other 24 hours on six nights (1938 January 19–25). The equivalent exposure is therefore forty-two hours – and the sky background is still black! Have these faint nebulosities been recaptured with modern photographic materials, which are so sensitive to fogging?

The Berthiot objective (120-mm diameter, f/10) did not remain idle for the first two years at Forcalquier. After several attempts, it captured the first *colour* photograph of the Green Flash, on a Lumière Autochrome plate. It was reproduced in black-and-white in *L'Astronomie*, 1932, page 90, illustrating an article by Emile Touchet. Predominantly green starch grains can be seen in the largest plume: the exposure was 0.1 second.

After 1938, the death of his wife and then the war cut de Kerolyr off. I heard nothing of him until 1946, when I received a telegram pleading for me to sell the equatorial mounting back to him. He visited me some days later in Marsillargues (Hérault), accompanied by Dr Marc Badel from Valence. The latter knew (via the S.A.F.) de Kerolyr's reputation, and had visited the "Paris Out-station" at Forcalquier. He had wanted an observatory of his own, and he gave Marcel de Kerolyr the job of bringing it to fruition. The equatorial mount and the Berthiot and Derogy cameras, and some other visual instruments, are reproduced in *L'Astronomie*, 1947 April.

Silence then falls again. Marc Badel told me that de Kerolyr had been in hospital and then in a rest home on the Côte d'Azur. We believe that he died there in 1969.

A sad end for an astonishing and very colourful person.

The Société Astronomique de France gave him the following awards: in 1933, the first Dorothea Roberts-Klumpke Award, only just instigated; in 1935, the Camille Flammarion Award; and in 1937, the Dorothea Roberts-Klumpke Award for the second time.

So the Société Astronomique de France honoured Marcel de Kerolyr. We may even say that it adored him; he was compared, and quite rightly, with the finest American astrophotographers, who had far superior instrumentation. And then it forgot him. *L'Astronomie* never mentioned him again, and never even published an obituary.

References

Fehrenbach, Ch., "La création de l'Observatoire de Haute Provence" in *Histoire et Avenir de l'OHP*, OHP, 1987. [De Kerolyr's equipment was illustrated in *L'Astronomie*, 1928 October, and the Couder-Prin telescope in 1933 September. Photographs taken by the early cameras were published in the issues for 1928 October, 1929 May and 1932 September, while those using the 81-cm appeared at intervals from 1933 May to 1938 April. – Eds]

Historical Session: The Contribution of Amateurs to Astronomy, Yesterday and Today

Concluding Remarks

André Koeckelenbergh
Vice-President, Société Royale Belge d'Astronomie, Méteorologie et Physique du globe, Observatoire Royale de Belgique
Avenue circulaire 3, B-1180 Brussels, Belgium

As I am given the occasion of closing this session, I must not fail to thank and to congratulate Madame Levasseur-Regourd, who presided over the meeting held yesterday. Neither must I fail to give due homage to the amateurs who have made astronomy's history. They are everywhere, and this Colloquium has shown this unequivocally. There are also those who are making astronomy's history today. The marvellous images of the heavens obtained by de Kerolyr delighted me when young, and the long conversations with Boyer and Focas, at Meudon, enriched my experience as an adult.

It would be ingratitude on my part not to recall the meticulous, constant, and disciplined work of some tens of sunspot observers, who send their observations every month to the "Centre de donnés pour les indices de l'activité solaire" (S.I.D.C.). They allow the Wolf number to be rapidly published and passed to organisations concerned with telecommunications, geophysics and space research. This is a field where the simple practicalities of the discipline forces amateur-professional collaboration. Another aspect of this collaboration is participation in expeditions to observe total solar eclipses. Past experience with such events shows that the self-denial and devotion of amateurs are such that many of them are so absorbed in technical tasks, crouching over chromometers and electronics that they often have no chance to view totality and the ephemeral splendour of the corona, which has been the very reason for their having frequently travelled thousands of kilometres, at their own expense!

This is history in the making. I am grateful to all the authors of the notable contributions that we have listened to with such interest over these two days. I also feel the same gratitude towards those who have enabled us to share in this precious common cultural heritage.

Part II

Observational Methods

Amateur Astronomers and the IAU Central Bureau for Astronomical Telegrams and Minor Planet Center

Brian G. Marsden
Harvard-Smithsonian Center for Astrophysics, 60 Garden Street, Cambridge, MA 02138, U.S.A.

Introduction

Of all the sections of the International Astronomical Union the Central Bureau for Astronomical Telegrams is undoubtedly the one that most concerns amateur astronomers. Just about anybody in the world with at least some familiarity with the sky has the potential to discover (or to think he or she has discovered) a comet or nova. If the object is real and sufficiently bright, it is very probably already known. Somebody has to be the first discoverer of every comet or nova, however, and soon after the IAU was established in 1919 it set up the Central Bureau to receive and to disseminate to the astronomical community news of such discoveries. Discoveries of supernovae in other galaxies, natural satellites of the planets, erupting x-ray sources and transient features on the planets are also dealt with by the Central Bureau, which since 1965 has operated at the Smithsonian Astrophysical Observatory in Cambridge, Massachusetts. The Central Bureau handles unusual minor planets in the vicinity of the earth, although the thousand or more ordinary minor planets routinely discovered each year (and with which amateurs are being increasingly involved) are more appropriately the province of the Minor Planet Center, set up by the IAU in 1947 and since 1978 also operated at the Smithsonian Astrophysical Observatory. About one-quarter of the subscribers to the various services of the Central Bureau and/or the Minor Planet Center are individual amateur astronomers or organizations of amateurs.

Discoveries

Comets. Throughout its history the dominant function of the Central Bureau has been the announcement of discoveries of comets. Both professional and amateur astronomers are involved in this activity, and it is perhaps of some interest to examine the relative contributions of these groups. The upper curve in Fig. 1 shows the total number of new comets discovered each year during the two-thirds of a century since the IAU was founded, and the lower curve shows the number of discoveries that were made by amateurs.

To eliminate the most dramatic variations from one year to the next and the indeterminacy that would otherwise occasionally arise in the computation of the relative amateur contribution, the plot actually gives annual values smoothed over

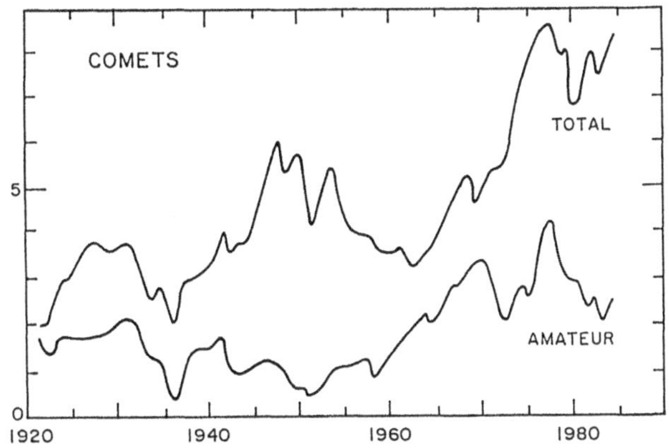

Fig. 1. The upper curve shows the total number of comets discovered per year, the lower curve the number found by amateurs. Values are smoothed over 5-year spans.

five-year spans. Accidental rediscoveries of periodic comets are included, but comets for which orbits could not be computed or which were not found until some years after the records of them were actually obtained are not. A comet can be credited to as many as three discoverers, and it is classified here as an amateur discovery if at least one of them was an amateur. With these definitions, it appears that, on the average, two new comets out of every five have been amateur discoveries.

Figure 1 shows that – by chance – very few comets were being discovered around the time that the IAU was established and that – again by chance – almost all of them were amateur discoveries. Although the professional contribution then quickly increased, the rate of amateur discoveries remained steady into the 1930s, mainly due to the efforts of a number of observers in South Africa. Following a general decline during the mid-1930s, both amateur and professional activity picked up during World War II. Immediately after the war the professional searches at the Skalnate Pleso Observatory in Czechoslovakia and the discoveries made during professional sky surveys in California dominated the scene, with amateur finds then at a very low level; there were no amateur comets at all for 3.5 years beginning at the end of 1948. The extensive Japanese amateur discoveries of the 1960s narrowed the gap, and while continuing Japanese comets and discoveries by the Australian amateur Bradfield continued to make an impressive showing during the 1970s, there was also then unprecedented professional activity with large Schmidt telescopes in both hemispheres. This professional activity, supplemented by the results from the Infrared Astronomy Satellite, has again been evident during the 1980s, with the result that amateur comet hunters now find two or three comets were year and are responsible for about one discovery out of three.

During the interval under consideration there have been 15 amateur astronomers who have discovered (or co-discovered) three or more comets. These most successful hunters are listed by country in Table I. Curiously, these individuals represent only six different countries, the total contributions from which are also shown in the

Table I. Records for amateur discoveries of comets

Country	%	TOT	IND	
Japan	35	61	27	Honda 12, Seki 6, Ikeya 5, Fujikawa 5, Sato 4, Tago 3
U.S.A.	22	39	21	Peltier 10, Friend 3, Machholz 3
R.S.A.	13	23	9	Skjellerup 5, Reid 5, Forbes 4
Australia	10	17	6	Bradfield 12
U.K.	3	6	2	Alcock 5
Canada	2	4	1	Meier 4

The TOTal number of discoveries in each country represents a percentage % of the whole and is divided among the number of INDividual observers shown.

table. There are only two additional countries from which more than three amateur discoveries have been made – namely, New Zealand and the U.S.S.R., each with five discoveries. The discoverers of the remaining 14 comets were located in seven other countries.

It is sometimes very difficult to decide whether a particular astronomer is an amateur or a professional. Of course, several individuals clearly make their first contributions as amateurs and their later ones in professional positions on the staff of an observatory. For the purposes of this paper, students participating in professional programs have been counted as professionals, but former professional astronomers who have formally retired but who continue to work with their own equipment and no specific financial support are regarded as amateurs. When it comes to considering discoveries made by astronomers there is even more of a problem. Few professional astronomers have ever actually been paid to discover comets, but many amateurs have received financial rewards for doing so! Amateurs may tend to observe visually, make searches specifically for comets, usually relatively bright ones at quite small elongations from the sun; while professionals tend to observe photographically and may by accident pick up comets, usually faint ones in the general vicinity of opposition. This is far from a generalization, however. The professional program at Skalnate Pleso was entirely a visual one, and amateurs found two comets photographically in late 1986. That these should be the first amateur photographic discoveries in fourteen years may seem surprising when one considers that films with fuzzy blobs on them are submitted to the Central Bureau by hopeful amateurs a couple of times each month. A typical month will tend to yield a somewhat larger number of alleged visual discoveries that can generally be explained as ghost images of planets and bright stars or close configurations near the limit of the telescope of stars that are individually too faint to be recorded on the particular charts used by the observers.

Novae. Nova hunting is often coupled with comet hunting as a suitable activity for amateur astronomers, but until 20 years ago the only novae found by amateurs were ones that were conspicuous to the unaided eye. Since the IAU's beginnings about

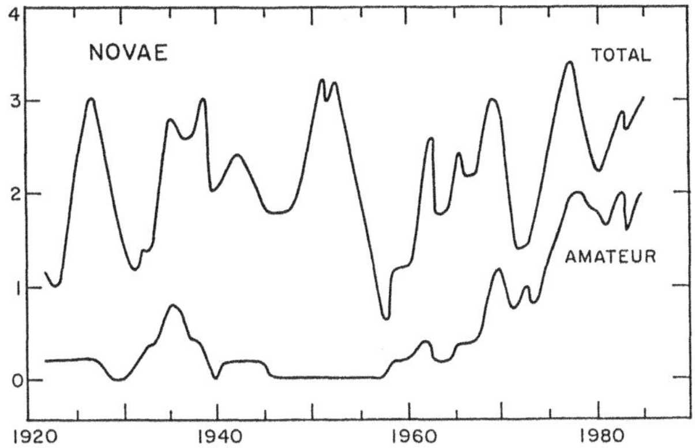

Fig. 2. The upper curve shows the total number of novae discovered per year, the lower curve the number found by amateurs. Values are smoothed over 5-year spans.

one nova in four has been discovered by an amateur. Annual rates, again smoothed over five years, are shown in Fig. 2.

Through the 1940s the occasional amateur discovery of a bright nova was of considerable interest to professional astrophysicists, but such finds represented a small fraction of the total novae found, as faint ones were noticed by professionals scanning patrol plates, often many years after the actual outbursts occurred. There were no bright novae between 1942 and 1960, and during the 1950s professional patrols effectively ground to a halt. Activity increased during the 1960s, and serious amateur searches got underway visually in the U.K. and photographically in Japan. For more than a decade there have been roughly two amateur nova discoveries and one professional nova discovery each year, although there is still always the possibility that photographic discoveries will surface years after the films were obtained.

Unlike comets, nova discoveries are normally credited to a single individual, and Table II shows the four amateurs who have made more than one discovery. The six remaining amateur discoveries are distributed among observers in five other countries. The first-discoverer rule is perhaps a little unfair. If it is relaxed, the U.S. would also be listed with two entries each for visual hunters Peltier and Collins.

Table II. Records for amateur discoveries of novae

Country	%	TOT	IND	
Japan	66	25	8	Honda 10, Kuwano 6, Wakuda 4
U.K.	18	7	4	Alcock 4

Visual nova hunting becomes very difficult when it approaches and exceeds the naked-eye limit, for it essentially requires the observer to commit to memory whole star fields in regions of the sky near the galactic equator. For reasons that are not altogether clear amateur photographic hunting for novae has so far had no success whatsoever outside Japan. Perhaps more so than in the case of comet discoverers, the problem of distinguishing amateurs from professionals again exists. Liller and McNaught are classed as professionals, even though the four novae found by the former during the past four years in Chile and the two found by the latter during the past seven months in Australia clearly represent labors of love that define these astronomers as amateurs in the truest sense.

There is also a more troublesome problem involving nova discoveries. When a comet suspect is bright enough to be found by an amateur and is reported reasonably quickly, it is generally confirmed (or otherwise) as a real comet by its position and physical appearance and by its motion in the sky during the days following discovery. Confirmation of a nova is more difficult. Checks are routinely made at the Central Bureau as to whether a nova suspect is a known major or minor planet, possibly picked up confusingly near a stationary point, and limited checking of known or suspected variable stars is also possible. But a lot of unknown and unsuspected variables are relatively bright, particularly in the southern hemisphere, and definitive classification as a nova may require a fairly extensive spectroscopic study, perhaps coupled with the careful examination of many exposures made of the field since the beginning of the century. At least in the past, professionals have had to be brought in to do this, and even those who are particularly interested in eruptive variables are usually occupied with other work or lack the telescope time or plate libraries that are needed. Very few professionals are willing or able to make the effort necessary to establish whether even an object as bright as eighth magnitude is really a nova or not, particularly if the object is not brightening.

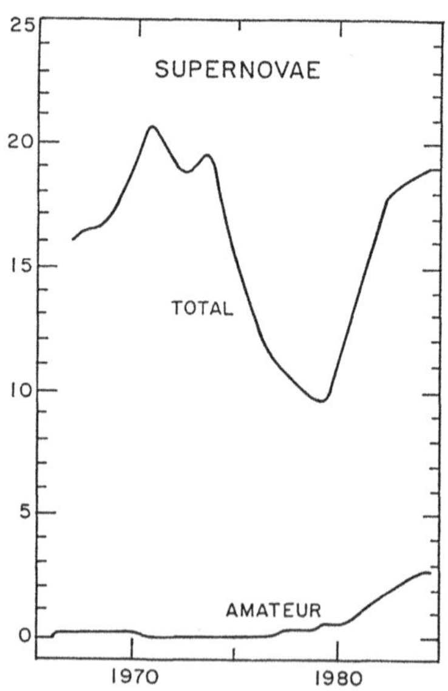

Fig. 3. The upper curve shows the total number of supernovae discovered per year, the lower curve the number found by amateurs. Values are smoothed over 5-year spans.

Supernovae. Although records of supernovae in other galaxies go back slightly more than a century, interest in them as phenomena distinct from ordinary novae dates back only half a century. The first amateur discovery of a supernova was by Bennett in South Africa in 1969, the second in Johnson in the U.S. in 1979. The smoothed totals in Fig. 3 show that professional discoveries had climbed to about 20 per year by the early 1970s, and the subsequent decline was due to the termination of the survey with the 1.2-m Schmidt at Palomar. Since 1980 there has been renewed interest and success by both professionals and amateurs. The annual total is back almost to the 1970 level, but with now some three discoveries per year by amateurs.
Amateurs who have found two or more supernovae are shown in Table III. The two Okazaki supernovae were also independently found by Evans. The truly remarkable contribution by Evans is entirely visual, while the Japanese discoveries are photographic. The single supernova 1987A in the LMC was independently found by the amateur Jones in New Zealand. Although supernova hunting has been widely discussed and indeed practised as an amateur activity in recent years, it is not as easy as the novice might think, and there are clearly many pitfalls for the unwary. As with visual nova hunting, the memory plays an important role in supernova hunting. Astigmatism can be a problem when galaxies are observed at different orientations in the sky. Comparison photographs of the galaxies to be scanned are generally useless, although there is some promise in the use of special drawings that attempt to depict the visual appearance of the galaxies. Photographic supernova searches require comparison with a library of earlier photographs of the same galaxies taken with the same telescope, emulsion, exposure time and sky conditions.

Table III. Records for amateur discoveries of supernovae

Country	%	TOT	IND	
Australia	68	15	1	Evans 15
Japan	18	4	2	Okazaki 2, Horiguchi 2

Minor planets. Although hunting for new minor planets enjoyed some popularity among amateur astronomers during the nineteenth century, and the U.S. amateur Metcalf and Austrian amateur Palisa were very active during the early part of the twentieth, there was a great hiatus in amateur discoveries between Palisa's last minor planet in 1924 and the Japanese amateur Urata's first in 1978. Although even the last of Palisa's 124 discoveries were visual, photographic patrols for these bodies had proven to be extremely efficient as soon as they were introduced in 1891, with several new discoveries possible on a single photographic plate. Just as amateurs were discouraged for several decades by professional successes in supernova hunting, so were they discouraged in the case of minor planets. From 1978 on, however, there has been a more-or-less steady increase in amateur activity (all photographic) with minor planets, and the rise from 2 in 1978 to 48 in 1986 is illustrated in Fig. 4. Unlike Figs. 1–3, Fig. 4 is on a logarithmic scale, and there has not been any smoothing. It should be noted that the numbers refer to the minor planets given provisional

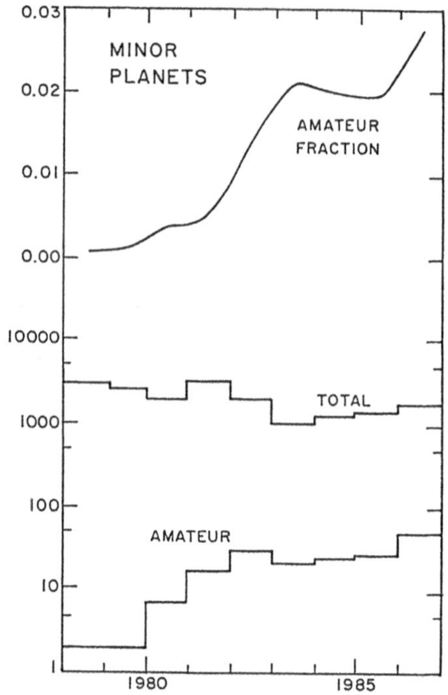

Fig. 4. The upper histogram shows (on a logarithmic scale) the total number of minor planets discovered per year, the lower histogram the number found by amateurs. The curve at the top shows the fraction discovered by amateurs.

designations, i.e., to the discovery or supposed discovery of an unidentifiable minor planet reported on the basis of the accurate measurement of the position of what may sometimes be only a single image. The fractional amateur contribution, rising from less than 0.001 to almost 0.03, is shown in the top part of Fig. 4. This fraction was actually as high as 0.075 (a total of 25 amateur discoveries) during the first third of 1987, although the fraction for the year will undoubtedly decrease in the future as professionals tend to be slower to process and submit their results than amateurs, and the prevailing weather patterns in the northern hemisphere mean that the largest professional minor-planet programs are most productive during the last third of each year.

Table IV shows all the amateur astronomers who have – from 1978 onward – discovered more than ten minor planets. All the amateur discoveries were made

Table IV. Records for amateur discoveries of minor planets

Country	%	TOT	IND	
Japan	81	160	11	Urata 85, Seki 62, Suzuki 40, Niijima 24, Furuta 11
Italy	19	37	3	Colombini et al. 26

in Japan or Italy, and the totals shown for the countries represent individual minor planets. The totals shown for the individuals involves some duplication, however, for the leading discoverer Urata works in colloboration with Suzuki, Niijima and several other amateurs not listed here.

Of the 197 amateur discoveries 25 are the principal discovery apparitions of numbered minor planets, 20 more are identified with numbered minor planets, and 45 are identified with unnumbered minor planets at other oppositions. Principal-apparition amateur discoverers of two or more numbered minor planets are shown in Table V, subject to the other qualifications noted in connection with Table IV. In addition, there are 42 discoveries for which preliminary orbits are available and 12 more observed on two different nights. What might be termed the "relative usefulness" of the amateur discoveries, whether measured in terms of the high fraction of principal-apparition discoveries of numbered minor planets (0.13) or the small fraction of discoveries observed on only a single night (0.26), is considerably greater than for professional discoveries.

Table V. Records for amateur principal-apparition discoveries of numbered minor planets

Country	%	TOT	IND	
Japan	92	23	8	Seki 11, Urata 9, Suzuki 3, Niijima 2, Furuta 2
Italy	8	2	1	Colombini et al. 2

Other Discoveries. Amateur discoveries in other areas of observational astronomy are understandably rather meager. Confirmed discoveries are generally restricted to occasional unusual variable stars found in the course of nova searches. Nova recurrences take place infrequently enough that, although it is in principle necessary for an observer only to monitor a few known points in the sky, the definite detection and intelligent announcement of an important outburst is unexpected enough that it effectively counts as a new discovery. Among the amateur contributors in this area Peltier in the U.S., Alcock in the U.K. and Jones in New Zealand are noteworthy. The occasional discovery of an unusual new feature on a major planet is in the same category, although the planets are probably better monitored nowadays by professional planetary patrols. Perhaps the most celebrated discovery of this type to be announced in the IAU Circulars was the white spot found on Saturn in 1933 by the British amateur astronomer and professional comedian Will Hay.

Recoveries of periodic comets as they return to perihelion have generally been the province of the professional astronomer armed with a large telescope and an accurate prediction. Predictions are sometimes significantly in error, however, particularly for a comet making its first predicted return or when the comet's motion is affected by large and irregular nongravitational forces. Since the IAU was established amateur astronomers have been involved with about one recovery in ten. Amateurs who have recovered or co-recovered more than one comet are shown in Table VI. There have also been three amateur recoveries in each of the U.K. and Australia (in the latter case all for the same comet) and one (by Jones) in New Zealand. Except

Table VI. Records for amateur recoverers of periodic comets

Country	%	TOT	IND	
Japan	52	15	4	Seki 12
U.S.A.	14	4	3	McClure 2
R.S.A.	10	3	2	Reid 2

for the faint comets recovered by Seki and McClure, almost all amateur recoveries have been visual. It has happened twice during the past three years that recoveries either were not made or could not be made until a comet was bright enough to be detected visually with a small telescope.

Follow-up Observations

Astrometry. Amateur astronomers are nowadays making an important contribution to the astrometric observations that follow a new discovery. Although highly desirable for new novae and supernovae, astrometry is of course essential if an orbit is to be calculated for a new comet or minor planet. A necessity during the first few days after a discovery, some astrometric coverage is useful as long as the comet or minor

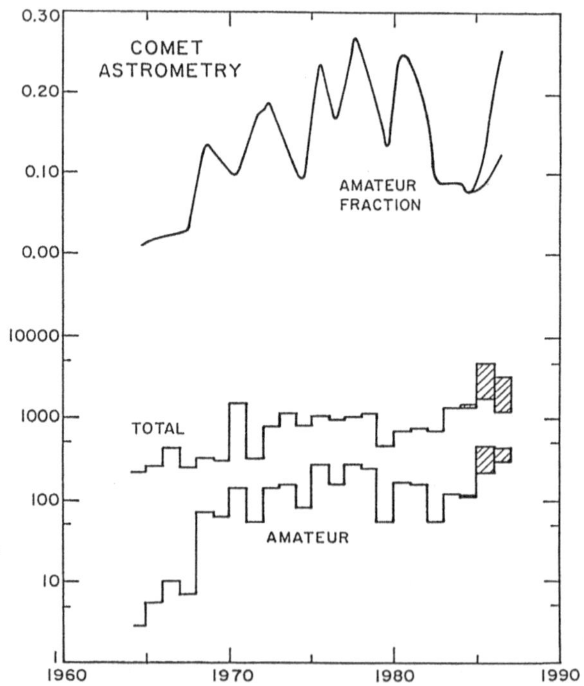

Fig. 5. The upper histogram shows (on a logarithmic scale) the total number of astrometric observations made of comets per year, the lower histogram the number made by amateurs. The shaded section refers to the observations of Halley's Comet. The curve at the top shows the fraction of observations made by amateurs, the upper and lower sections following the bifurcation being according as to whether Halley's Comet is excluded or included.

planet can be observed, particularly if there is some thought of recovering the object at subsequent apparitions.

Figure 5 illustrates the situation with regard to astrometric observations (essentially all photographic) of comets since 1964, the first year for which the machine-readable records of the Central Bureau/Minor Planet Center are essentially complete. As in Fig. 4, the scale in the lower section is logarithmic, and there has been no smoothing. The shaded section from 1984 onward refers to the observations only of Halley's Comet, to which the attention given has been completely out of proportion and largely useless. Up to 1967 the only amateur astrometric observations were a handful each year by Waterfield in the U.K. The top part of Fig. 5 shows that the fractional amateur contribution has averaged one out of seven but has been as high as one out of four. In the section from 1984 onward the upper curve is that when Halley's Comet is omitted, the lower curve that when it is included. Amateurs can take some consolation in the fact that the "band-waggon" effect of this well-publicized object was significantly less for them than it was in the case of the professionals.

Table VII lists the ten amateur observers who have made at least 100 cometary astrometric observations. In considering the Japanese results it should also be noted that Seki measured the films by Kojima, and Urata measured those by Suzuki and Furuta. The five countries tabulated are followed by France (with 89 observations), West Germany (80), the United States (50), and seven others (43 altogether).

Table VII. Records for amateur astrometric observations

Country	%	TOT	IND	
Japan	60	2168	39	Seki 973, Urata 232, Suzuki 158, Furuta 128, Kojima 122
U.K.	14	525	17	Manning 164, Waterfield 150
Australia	9	336	1	Herald 336
Italy	6	222	5	Colombini et al. 123
R.S.A.	3	100	1	Hers 100

In general, the accuracy of the cometary astrometric observations by amateurs compares very favorably with that of observations by professionals around 2 arcsec. An observation by an amateur is about equally likely as that by a professional to be affected by some gross error, more often than not one of exactly an hour or exactly a day in the indicated time of mid-exposure. The speed with which amateurs reduce and report their data is generally rather greater than that of professionals.

The statistics for astrometric observations of minor planets are not illustrated here. Several of the comet observers (though not those in the U.K. or R.S.A.) also routinely make observations of unnumbered minor planets, and if observations of the brighter numbered minor planets are also included, the contribution from several observers in West Germany and one observer in East Germany can be noted. While the accuracy of amateur minor-planet astrometry is comparable to that of amateur cometary astrometry, the best professional data are internally consistent to better than 1 arcsec. This is particularly true of CCD data, and with the anticipated improvements in reference-star positions much of the absolute professional astrometry will be noticeably more accurate than the amateur data.

Photometry. By far the most pervasive contribution by amateur astronomers to the IAU Circulars nowadays concerns photometric data on comets and novae, and occasionally also on other unusual variables, notably information on the fadings of variables of R CrB type. Although some of the non-cometary observations are obtained photoelectrically, most of them consist of comparisons by eye with reference stars nearby. The visual magnitude estimates that are published are only a tiny fraction of all that are coordinated, made and eventually collected by organizations of variable-star observers in several different countries, and the following speaker will discuss such observations in detail.

The cometary magnitude estimates published are also only a small fraction of such data handled by similar national groups, as well as by the International Comet Quarterly and the amateur net of the International Halley Watch, the latter to be discussed by another speaker. The comet magnitudes refer to the total coma and are sometimes accompanied by estimates of coma size, tail length and other physical characteristics of the comet. The selection of amateur photometric data published on the IAU Circulars is intended to be as current as possible, so that professional observers can appropriately plan their own work.

Orbit Computations

Although some may consider that the incorporation of remarks on orbit computations by amateurs is inappropriate in a part of this Colloquium dealing with observations, such computations are intimately involved with astrometric observations and in some sense are only a minor extension of them. The computations range from that of a very preliminary orbit of a new comet or minor planet, through successive improvements of the orbit, to the point where it is necessary to consider the effects of planetary perturbations and to make predictions for future potential apparitions. Computations for comets might involve some allowance for non-gravitational effects, and those on minor planets a search for past identifications of the same object. Complementing their involvement with astrometric observations directly, several amateur astronomers are nowadays doing fine work in all these areas, and it is being published by the Minor Planet Center.

The British Astronomical Association has the longest tradition for such work, dating back to the 1920s, particularly with respect to predictions for the returns of comets, although preliminary orbits were also sometimes calculated. This author started out as an amateur involved with such work in those far-off days of logarithms and mechanical calculating machines. In Japan and in West Germany the amateur computations 30 or 40 years years ago were more likely to involve minor planets, and a good many preliminary elliptical and circular orbits were produced. One of the Japanese orbit computers of that time was the great observer Seki, while his countryman Hasegawa, who will be speaking later in this session, was making many computations of preliminary orbits of comets. One of the German amateurs, Kippes, embarked on the subject of minor-planet identification, work that is in practice more difficult because of the need to organize databases of both astrometric observations and preliminary orbits. Kippes did this work by hand, as he still does four decades later.

For many amateurs, however, the availability of modern microcomputers has revolutionized orbit-computational work, just as main-frame computers (as well as microcomputers) have done so for professionals. The amateurs in the British Astronomical Association, notably Milbourn, who has also done his share of hand computation, continue to work on comet orbits, and German (Landgraf, Kretlow) computations on minor-planet orbits and identifications have been supplemented by those from amateurs in Belgium (Goffin) and Canada (Lowe). But the greatest and most impressive activity is by the Japanese amateurs, who tackle minor-planet and comet orbits with equal facility, several of them utilizing and exploring new techniques for finding minor-planet identifications (Nakano, Kobayashi, Urata, Furuta, Oishi) and solving for cometary nongravitational parameters (Nakano, Kobayashi).

Concluding Remarks

The most obvious conclusion that can be drawn from this paper is that Japanese amateur astronomers are collectively the best in the world and that they completely dominate in all the areas considered except that of the discovery of supernovae, where they defer to Australia. In addition to its lone supernova discoverer Australia's general very high standing is due also to one outstanding comet hunter and one extremely productive amateur astrometrist. Amateurs in the United Kingdom make a very strong showing in astrometry, and – thanks basically to one individual – also in visual nova and to a lesser extent comet hunting. The only other countries showing good collective effort in multiple areas are the United States and South Africa. There are some contributions from France, West Germany, Italy, Belgium and New Zealand but relatively little from Second World countries and none from countries of the Third and Fourth Worlds.

The intense Japanese and British contributions to observational astronomy are curious because these countries are both notorious for their cloudy skies. The other three leading countries are of course geographically larger and more blessed with good weather, but the United States also has a considerably larger population, and for that reason one must say that its total contribution is disappointingly small. Part of this is undoubtedly because U.S. amateurs tend to shun anything that may involve anything mathematical (in spite of the widespread availability of modern microcomputers); all 50 astrometric positions of comets (as well as all the positions of minor planets) were made by one largely unknown amateur, Terry Handley of Burlington, New Jersey; and the only orbital work has been the suggestion of a number of minor-planet identifications by another relative unknown, Frank Bowman of Cincinnati, Ohio.

Tsutomu Seki of Kochi, Japan, emerges quite definitely as the leading all-round amateur astronomer in the world, his astounding productivity ranging from the visual discovery of comets to photographic astrometry (including comet recovery) and occasional orbit computations. Honda and Alcock, and to a lesser extent Peltier, Reid and Jones, appear as outstanding observers involving both comets and novae. Urata, and to a lesser extent Furuta, contribute to both astrometric observations and minor-planet orbit computations and identifications. Essentially all the amateur contributions to the Central Bureau/Minor Planet Center have been made by fewer than a hundred individuals, not one of whom is female.

What of the future? The straightforward answer is "More of the same", and to a large extent this will undoubtedly be true. Professionals have talked about making automated surveys, particularly for supernovae, using devices involving scanning CCDs, but the productivity to date has not exactly been impressive. A few rugged individualist amateurs bounced back from the threat of the photographic plate, and they will do the same in the case of the CCD. And just as some amateurs joined the professionals and have undertaken photographic survey work of their own, others can in the future be expected successfully to exploit the use of the CCD. Visual comet hunters still rule the productive twilit skies near the sun, but one suspects that other hunting techniques, in the hands of both amateurs and professionals, are likely to encroach upon this territory in the future.

Discoverers are going to have to be aware that amateur astronomers will need to play a greater role in obtaining follow-up observations. Amateur astrometrists are already performing well in getting positions of new comets, but as professionals do less and less of this important work the burden on amateurs will have to increase. The nova situation is even more serious, for if hunters are going to continue to produce nova suspects that are well below naked-eye visibility, it will become necessary for amateur spectroscopists, not only to obtain good-quality spectra of the suspects, but also correctly to interpret these spectra. This is probably in fact the single most important problem for observational amateur astronomers to address in the immediate future.

Contributions of Amateur Astronomers to Variable-Star Observing

Janet Akyuz Mattei
Americal Association of Variable Star Observers (AAVSO), 25 Birch Street, Cambridge, MA 02138, U.S.A.

Astronomy is a unique field of science in which amateur astronomers have made and continue to make significant contributions.

One major area of the science of astronomy is the study of variable stars. There are more than 28,000 known and catalogued stars that change in brightness – variable stars – and about 15,000 suspected to be variable. These stars need continuous and systematic observing over decades to determine their behaviour and to record any of their unusual or rare activity. However, there are not enough professional astronomers or telescopes to observe these stars regularly. Therefore, variable star astronomy needs amateur astronomers to be the record keepers for these stars. It was for this reason that as early as 1844 and continuing throughout the second half of the nineteenth and early twentieth centuries, the astronomers F.W. Argelander of Germany, Sir John Herschel and J. Baxendell of England, D.F.J. Arago of France, and E.C. Pickering of the United States all advocated systematic variable star observing for amateur astronomers. The encouraging appeals from these leading professional astronomers resulted in the formation of several organized groups of variable stars observers, first in England – the British Astronomical Association (1890), Variable Star Section; then followed by the American Association of Variable Star Observers (1911); the Association Française des Observateurs d'Etoiles Variables (1921); the Royal Astronomical Society of New Zealand, Variable Star Section (1927); and the Japanese Astronomical Study Association (1945). Today, there are about 25 variable star observer groups around the world.

Since 1970, with the advancement of technology, variable stars have gained special significance as observations by professional astronomers have extended from X-ray to radio wavelengths. During this time, the number, range, and importance of contributions from amateurs have increased dramatically.

I wish to share with you some specific examples of the significant contributions variable star observers have made in recent years. Although these are examples from AAVSO records, they are testimonies to variable star observers around the world, for more than 50% of the observations received annually by the AAVSO come from observers outside of the USA. In fact, France, South Africa, Hungary, Germany, Canada, and the Netherlands were the leading countries in contributing observations to AAVSO data files last year.

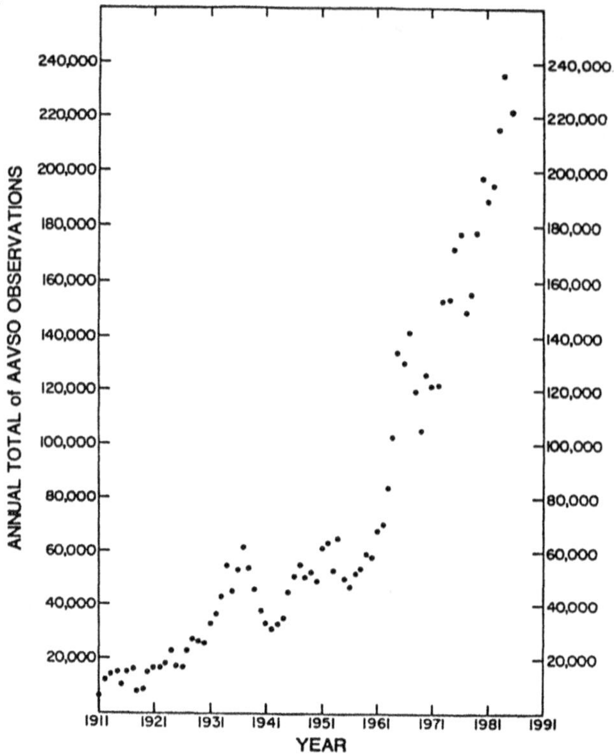

Fig. 1. Observations received each year by AAVSO, 1911–1986.

As interest in variable stars has grown in the professional community, the interest and the dedication of variable star observers also has grown. The number of observations received annually by the AAVSO has increased exponentially, as seen in Figure 1. The grand total of the observations archived by the AAVSO since 1911 is now over 5.5 million, contributed by more than 4000 observers around the world, Figure 2. In fact, the dedicated French variable star observer, Paul Vedrenne, made the 5,000,000 observation in the AAVSO data files.

In the 13 years that I have been the director of the AAVSO, we have supplied variable star observations made by amateur astronomers to over 1200 professional astronomers and researchers. Figure 3 is a histogram of the number of requests for data filled each year since 1974. These requests are for the following information :

1) to correlate data in multi-color wavelengths;

2) to schedule observing programs with large telescopes;

3) for simultaneous monitoring of scheduled stars;

4) for data analysis;

5) for reference materials to write articles on variable stars; and

6) for student science projects.

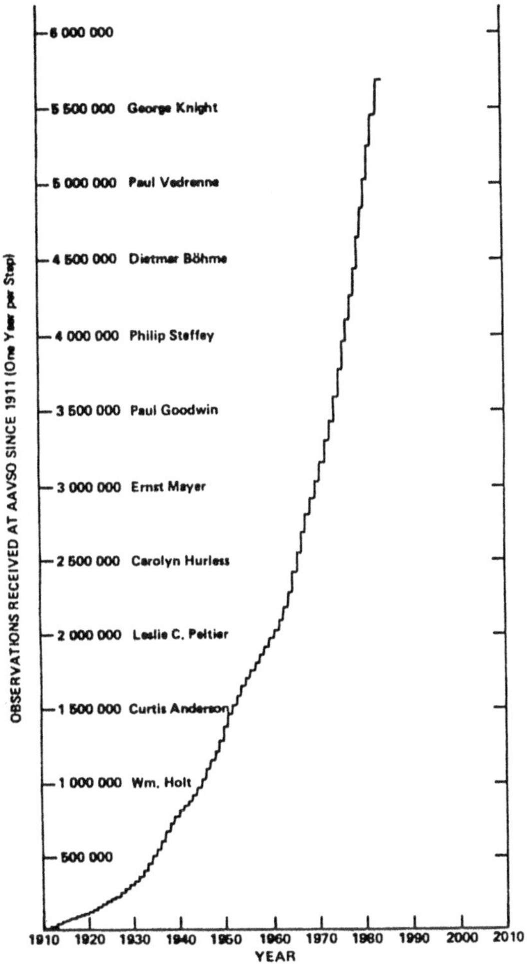

Fig. 2. Total of observations received by AAVSO, 1911–1986. Observers who made milestone observations (1,000,000; 1,500,000; etc.) are indicated on the graph.

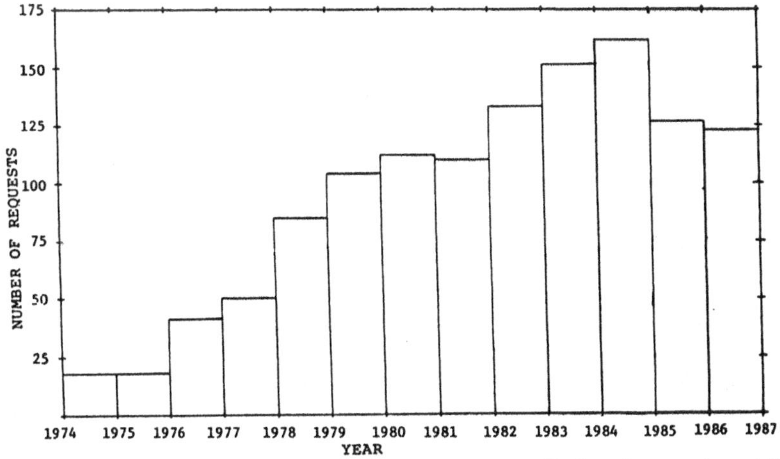

Fig. 3. Number of special requests for AAVSO data filled each year since 1974.

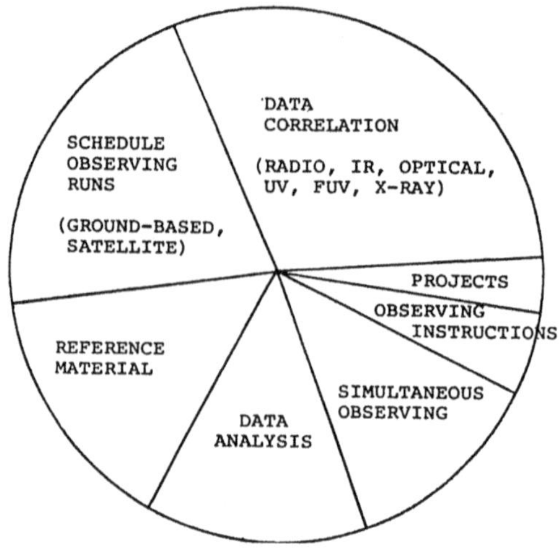

Fig. 4. Areas in which AAVSO data and services were used during 1984.

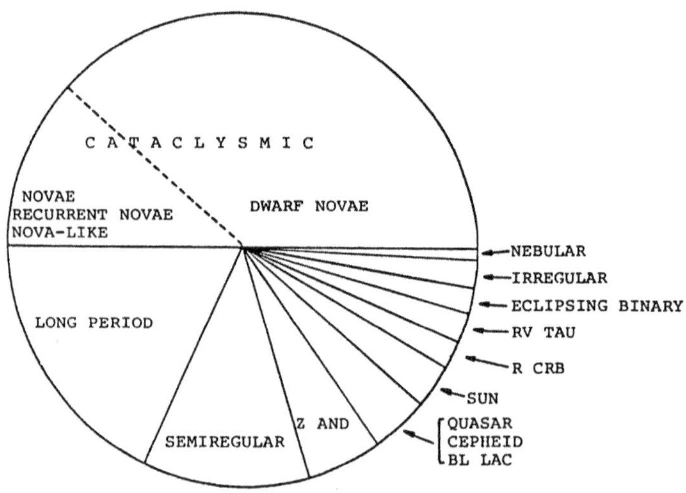

Fig. 5. Types of variable stars for which AASVO data were requested during 1984.

Figure 4 is a piegram showing in what areas the AAVSO observations were used in 1984. Observational information has been requested on all types of variable stars, but due to very active satellite research programs in recent years, the largest number of requests has been for observations on cataclysmic variables. Figure 5 is a piegram showing the different types of variable stars for which AAVSO observations were requested in 1984.

The most important and vital contributions of variable star observers in recent years have been :

1) assisting professional astronomers in scheduling observing programs with large telescopes or instruments aboard satellites;

2) providing simultaneous optical coverage and immediate notification of the behaviour of the scheduled stars during the observing runs; and

3) providing high quality optical observations for data correlation in multi-color wavelengths.

I would now like to expand on these contributions.

1. Scheduling observing programs

Continuous observations provided by amateur astronomers are essential to professional who wish to schedule observing programs with large earth-based telescopes or satellites equipped with specialized instruments during a specific phase of light variation of a variable star.

In the area of pulsating, long period variables, using both long-term and up- to-date observations provided by observers, together with AAVSO mean curves for each star, we predict the maxima and minima dates of 560 stars each year and publish this information in the AVVSO Bulletin. This publication is indispensable to both professional and amateur astronomers in setting up observing programs of long period variables. In addition to this, for specific observing programs we are able to predict the date and the brightness of a variable star at the requested phase, information which is crucial for the most efficient use of time on large telescopes.

We recently assisted astronomers at the University of California in this way. They were interested in observing with a radio telecope those bright long period variable stars known to be SiO masers, at maximum light. They called us to ask if the stars in their observing list would be bright at the time scheduled for observing. Extrapolating the AAVSO data, we found that all the stars on their list would be faint, below the limit of detection of the radio telescope, at the time the radio observations were scheduled. This crucial information enabled the astronomers to reschedule their observing time for a later date and thus make maximum use of the radio telescope.

In the area of close binary eruptive systems, usually referred to as cataclysmic variables, amateur astronomers have been playing a crucial role since 1973, when astronomers surveying the Cygnus Loop in X-ray wavelengths with sounding rockets accidentally detected ultra-soft X-ray emission from one of the most popular cataclysmic variables, SS Cygni. This important finding, predicted earlier theoretically, accelerated both earth-based and space research on these systems. Since the AAVSO observing program and data files contain a large number of these stars, professional astronomers have sought and continue to seek the help of variable stars observers in planning and scheduling almost every observing program of cataclysmic variables.

We had a very interesting and exciting collaboration recently with Dr. Janet Drew and her colleagues at Oxford University, England. They were interested in observing the eruption of the cataclysmic variable, YZ Cancri, in the ultraviolet with the IUE satellite. This star has frequent eruptions every 12 to 17 days that last one to three days. Superimposed on this behavior are bright and long superoutbursts that occur every two hundred days or so. To predict the occurrence of either of these eruptions is very difficult, indeed. The IUE satellite has to be scheduled months in advance. However, one can make the best guess as to when the next outburst will occur only about a month beforehand. The IUE run for YZ Cancri was scheduled for the seconf half of February, and astronomers were going to travel to the IUE station in Spain during the observations. January data on YZ Cancri indicated that the expected outburst most probably would take place not in the second hal of February but in early March. Using this crucial information, astronomers rescheduled their observing for early March. On March 1st, they went to Spain and waited to be alerted, depending on the observations of amateur astronomers. Observers all around the world were observing YZ Cancri very closely, and on the night of March 3, Gerald Duck, a dedicated observer in Massachusetts, reported that the explosion of this star had started. His detection was followed by observers around the world, such as Stephen Lubbock in England and Sei-ichi Sakuma in Japan. Throughout the ourburst, observers from around the world called to give us the most up-to-date information. YZ Cancri started to fade on March 5. The astronomers were able to obtain the earliest and most complete UV spectra of the short outburst of this star. This type of data is vital to understand the disk structure around the white dwarf on this close binary system, and the nature and cause of the outbursts. Since the duration of these outbursts is so short, to be able to schedule and observe this type of outburst is a real challenge and an astronomical victory only to be won with the help of dedicated variable star observers around the world.

2. Simultaneous optical coverage and immediate notification

During scheduled earth-based and satellite observing programs, astronomers again request variable star observers to provide simultaneous optical coverage of the scheduled stars and to inform them of the stars' behavior regularly. This simultaneous optical coverage has been crucial for the most effective use of telescope and instrument time, and for obtaining vital information during rare events of variable stars. It is very exciting to know that in this age of high technology, professional astronomers depend on amateurs in planning their observing, and in selecting the object at which to point their large earth-based telescopes or satellites. The dedication, the enthusiasm, and the high quality data of variable star observers have made this unique collaboration possible.

The following expert from a letter from an astronomer who has benefited from data provided by variable star observers demonstrates the pivotal role amateur astronomers play. Dr. Michael Bode from the University of Manchester, England, wrote :

"The AAVSO has several times recently provided me with data on the visual behavior of variable stars being observed at other wavelengths by satellite or ground-based facilities... In one classic case recently we obtained precise simultaneous observations of the dwarf nova SU

UMa with EXOSAT, IUE, and IRAS. Although several large ground-based observatories were prepared to make optical observations they were all 'clouded out'. The only point in this part of the spectrum was obtained by an AAVSO member... This is only one example of the many occasions that the AAVSO has provided invaluable assistance."

Amateur astronomers around the world have played a vital role in research on cataclysmic variables with X-ray and ultraviolet satellites. The observers have worked closely with Drs. France Cordova and John Middleditch of the Space Astronomy and Astrophysics Group at Los Alamos National Laboratory, and Keith Mason, of the Mullard Space Science Laboratory of the University of London, England. The fruitful collaboration between this group of astronomers and the AAVSO resulted in 23 scientific papers. Dr. Cordova wrote, "The AAVSO data were used to augment high energy X-ray and ultraviolet satellite data and were essential to the conclusions drawn in these papers."

Dr. Cordova and her colleagues enumerated some of the astrophysical research programs that they collaborated on with the AAVSO and described how the AAVSO helped them in their investigations using the X-ray and ultraviolet satellites :

"1. During the 6 month all-sky scanning phase of the X-ray astronomy satellite HEAO–1, AAVSO observers kept vigil over ~200 dwarf novae so that nearly simultaneous visual data would be obtained during the X-ray observations of these stars. The result was the first survey of the X-ray emission from a cataclysmic variable and the detection of X-ray emission from a few members of each cataclysmic variable subclass.

"2. During the pointing phase of HEAO–1, AAVSO observers alerted us to optical outbursts of dwarf novae so that these stars could be observed by the satellite during outbursts, when enhancements in the X-ray flux could be expected. The result was the discovery of soft X-ray pulsations from two dwarf novae (SS Cygni and U Geminorum); these were the first detection of soft X-ray pulsations from any astrophysical source.

"3. To clarify the nature of the soft X-ray pulsations we undertook a program of high time resolution optical photometry of dwarf novae using ground-based telescopes. AAVSO observers monitored a large sample of dwarf novae before and during our observing runs, and alerted us to incipient outburst activity. In this way we maximized our chances for detecting optical pulsations by observing only those dwarf novae undergoing outbursts. The results were the first multi-color spectra of dwarf nova pulsations.

"4. AAVSO observers alerted us to the outbursts of dwarf novae during our many observing runs using the International Ultraviolet Explorer (IUE) Satellite. The result was the accumulation of many UV spectra of dwarf novae in various optical outburst states. The AAVSO light curves of dwarf nova outbursts were essential in interpreting the UV data, and led to the discovery that the UV outburst is delayed with respect to the optical outburst.

"5. The HEAO–2 satellite (or "Einstein") with its much improved sensitivity over HEAO–1, discovered X-ray emission from about 70% of the 70 or more cataclysmic variables it observed. AAVSO observers monitored nearly all of these stars near the time of the X-ray observations. This information was vital in deriving a luminosity function for cataclysmic variables, and in testing theories for the high–energy emission in these compact systems.

"... The above examples represent only the results from our own collaborations with the AAVSO, and, as such, only a small part of the significant contribution AAVSO observers are making towards furthering basic research in astrophysics."

3. Data correlation

Variable star observations made by amateur astronomers have been used by professional astronomers to correlate polarimetric, photometric, and spectroscopic multi-wavelength data, ranging through radio, infrared, optical, ultraviolet, far ultraviolet, extreme ultraviolet, and X-ray wavelengths, obtained with earth-based large telescopes at observatories such as Mount Palomar, Kitt Peak, Cerro Tololo, Lick, McDonald, Mount Lemon, Whipple Multi-Mirror, Dominion Astrophysical, Tenerife, and Jodrell Bank, the Very Large Array Radio Telescope, and instruments aboard satellites such as the International Ultraviolet Explorer (IUE), High Energy Astronomical Observatories 1 and 2 (HEAO–1 and 2), Apollo–Soyuz, Voyager, Infrared Astronomical Satellite (IRAS), and the European X–ray Observatory Satellite (EXOSAT).

In the area of pulsating long period variables, observations made by amateur astronomers are essential both to theoretical astronomers working to understand the mechanism of the variation in these stars, and to observational astronomers who are observing these stars in optical, infrared, and radio wavelengths to obtain more information on the kinematics, chemistry, and evolution of the extended atmospheres of these stars.

Drs. L.A. Willson, G. Wallerstein, and C.A. Pilachowski, in their scientific paper entitled "Atmospheric Kinematics of High Velocity Long Period Variables ", in the Monthly Notices of the Royal Astronomical Society, 198, 483 - 516, used AAVSO observations and acknowledge the AAVSO in the following way :

"No paper on LPVs is complete without an acknowledgement to observations would be uninterpretable."

In the area of symbiotic variables, the data from variable star observers again played a crucial role in correlating observations obtained with radio telescopes. The star in question was the bright circumpolar variable CH Cygni, which has recently been reclassified as a symbiotic star consisting of a red giant and a hot blue component, imbedded in nebulosity. CH Cygni has two periods, a short one of 100 days, and another, longer period of 700 days. This variable star had been unusually bright since 1977. Both its short and long period variations were washed

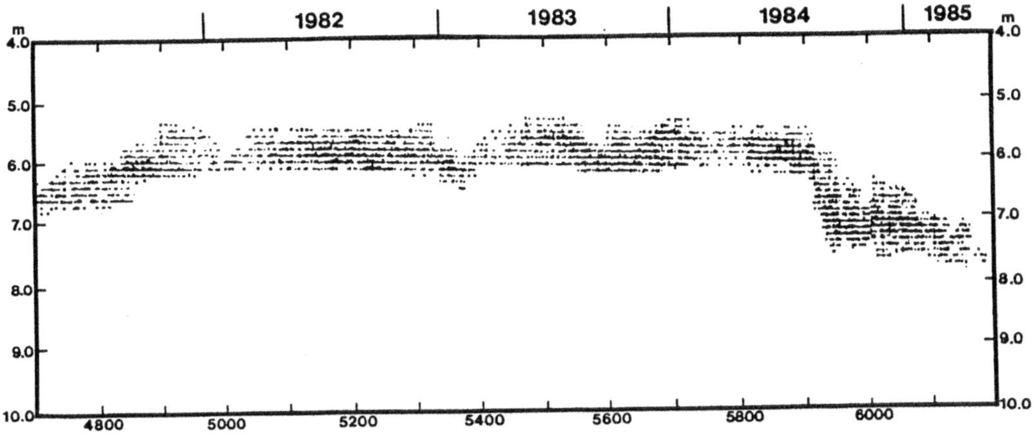

J.D. 2,440,000+

CH CYGNI

Fig. 6. AASVO visual light curve of CH Cygni from April to May 1985. Each dot represents one observation.

out. In June 1984 observers reported a sharp decline in its light variation, in which it faded from magnitude 5.5 to 7.5, Figure 6. During this time, Dr. R. Taylor from the University of Groningen, Netherlands, and Dr. E. Seaquist of University of Toronto, Canada, observing with the Very Large Array radio telescope, discovered a radio outburst and jets accompanied with increased flux density observed from April 1984 to April 1985. The increased flux density corresponded to the optical drop. Recently Dr. Seaquist wrote :

"...The AAVSO light curve for this star is playing a crucial role in the analysis of this event. What would we have done without the work of all the variable star observers ?"

The analysis of long-term visual data of cataclysmic variables has thus far revealed interesting correlations. There is, for example, a strong correlation among the duration of the outburst, rate of rise and decline from outburst, and the orbital period and the mass of the secondary component of the system. There is a relationship between the orbital period and the decline from outburst, in that the smaller the orbital period the faster is the decline from outburst. The orbital period of an eruptive close binary system is an important physical parameter that has a direct bearing on the mass of the secondary, the rate of mass transfer, the size of the system, and possibly the evolution of the system. To obtain orbital periods of cataclysmic variables spectroscopically requires large telescopes and is a difficult procedure. However, an indirect way to obtain the orbital period of a special type of cataclysmic variable – SU Ursae Majoris stars – is to observe the small-amplitude, periodic modulations (superhumps) during long outbursts (supermaxima) and to obtain their periods which are two to three percent longer than the orbital period of the system itself. Thus, if supermaxima of a system can be predicted, then through the observations of superhumps the orbital period of a SU UMa system may be determined.

Long-term analysis of the light curves of these types of stars has shown that one can predict the occurence of supermaxima (Mattei 1983) and in the recent years through the correlation of high speed photometric data and the AAVSO optical light curves six dwarf novae (RZ Sagittae, TY Piscium, HT Cassiopeiae, SW Ursae Majoris, VY Aquarii, and T Leonis) have been confirmed to belong to SU Ursae Majoris subclass, and orbital periods have been obtained for them.

Amateurs astronomers, through their valuable observations of variable stars, have provided complete, historical records on hundreds of variable stars, and, as we have seen, have played a major role in helping professional astronomers push back the frontiers of astronomy and gain understanding of a number of important and puzzling phenomena in astrophysics such as stellar pulsations, stellar instability, and time-varying processes in stars. The value of the contributions of variable star observers may be summarized by Dr. F. Cordova, who said, "The real discoveries of these stars belong to the men and women who keep vigil on them."

It is certainly both an honor and a pleasure for me, representing the AAVSO, to be joining in the celebrations of the 100th anniversary of the Société Astronomique de France. May the success of SAF and the valuable contribution of its members to astronomy continue for many centuries.

Observations of Comets, Minor Planets and Meteors in Japan

Ichiro Hasegawa
2-3-11, Saidaijinogami, Nara 631, Japan

1. Comets

In the early period of this century, some comets were discovered by Japanese amateur astronomers. Just fifty years ago, in 1937, Minoru Honda began to search for systematically comets with his 15-cm reflector. He observed Comet Encke without an ephemeris in November, 1937, and made his firts discovery of a comet in 1940. This comet was also discovered independently by Shigeki Okabayashi (Comet Okabayashi–Honda, 1940e). Subsequently, Honda discovered Comet Friend–Reese–Honda (1941a) in 1941. Unfortunately, Okabayashi was killed at the Second World War in 1945.

After the end of the war, Honda discovered ten comets during 1947 and 1968. In 1968, he discovered three comets in a year. He has been an active observer of comets and novae. Stimulated by Honda's success, Tsutomu Seki began his comet hunting in 1950, and his long and patient efforts were rewarded with the discovery of Comet Seki, 1961f. Seki later discovered six comets, and rediscovered six short-period comets before 1981. In that year, Seizo Goto, a well-known telescope maker in Japan, donated a 60-cm reflector. Since then, with the Goto telescope, Seki has made many astronomic observations of comets and minor planets. He has rediscovered some periodic comets and has discovered many minor planets every year.

Kaoru Ikeya discovered his first comet in 1963. In 1965, Ikeya and Seki independently discovered a sun-grazing comet. The interval between their discoveries was only 15 minutes. Comet Ikeya-Seki (1965f) had a great influence upon amateur astronomers, and observation and searching for comets became more active in Japan.

In the morning of May the 1st in 1968, six Japanese amateur observers discovered a 7–magnitude comet independently. It was named Comet Tago-Honda-Yamamoto (1968a) after three of the discoverers. At the present moment, there are seven comets named after three Japanese amateurs, and four after two Japanese comet hunters. It may be noteworthy that in 1975, Hiroaki Mori discovered two comets on the same morning.

All the observers mentioned above made visual discoveries. Since 1970, however, some observers have come to use photographic techniques to discover comets. Nobuhisa Kojima (1970r and 1972j), Toshihiko Ikemura (1975b), Takeshi Urata and Tsuneo Niijima (1986o) discovered their comets (fainter than 12th magnitude) by photographic observations.

Since 1971, comet observers in Japan gather once a year to have a comet conference. The convention held in March this year invited Dr B.G. Marsden, a comet expert from the Smithsonian Institution. It is estimated that more than fifty Japanese comet hunters are searching for comets every night. The statistics of comet discoveries are given in the accompanying table.

Comet Discovery Statistics (1961-1986)

Year	Discovery of New comets (Total)	Discovered by Amateurs (Breakdown)	Discovered by Japanese amateurs (Breakdown)
1961	3	2	1
1962	2	2	2
1963	5	2	1
1964	3	2	2
1965	5	2	1
1966	4	1	1
1967	4	3	2
1968	7	5	3
1969	5	3	2
1970	6	5	4
1971	1	1	1
1972	6	2	1
1973	8	0	0
1974	5	2	0
1975	13	8	5
1976	5	2	0
1977	8	1	0
1978	11	7	2
1979	6	3	0
1980	10	4	0
1981	7	0	0
1982	2	1	0
1983	13	2	2
1984	12	5	2
1985	6	1	0
1986	10	4	1
Total	167	70	33

2. Minor Planets

Recently, about seven or eight years ago, Japanese amateurs became able to use larger instruments to make photographic observations. They use telescopes larger than 30–cm, some 60–cm or greater. They can now observe fainter minor planets easily near opposition. The most successful discoverers of minor planets are Seki, Suzuki, Urata, Niijima and Kizawa.

Amateur orbit calculators also devote their efforts to determining the orbits of minor planets, and to identifying their perturbations, and have made fruitful results in confirming minor planets to be numbered at the Minor Planet Center of the IAU. Hideo Oishi, Syuichi Nakano, Takeshi Urata, Toshimasa Furuta and Takao Kobayashi are the most keen and successful computers. Nakano also determines comet orbits and makes predictions of short-period comets every year.

Facilities of the Tokyo Astronomical Observatory for communicating and confirmating discovery observations are most valuable and are encouraging the Japanese amateur activities.

3. Meteors

In 1920, the Oriental Astronomical Association was founded by Professor Issei Yamamoto (1889-1959). One of the most active fields of the OAA has been meteor observations. Kojiro Komaki (1903-1969) made his first observations of meteors in 1921, and led the observations of Leonid Meteors in 1933. His most important observation seems to be that of the great apparition of Lyrids in April, 1945. It occured at the end of the Second World War and because of the locality of meteor apparition, no other meteor observer witnessed this remarkable meteor shower.

Komaki organized the Nippon Meteor Society in 1956. This society is the most active group of the amateur astronomers in Japan. The Japanese meteor observers work with many kinds of techniques to observe meteors, such as the traditional visual observations, the telescopic and photographic ones. More than one hundred photographic meteors are observed by single or double stations, and their heliocentric orbits determined, in a year. Recently, the association of the Monocerotid meteors in December with Periodic Comet Mellish (1917 I) was confirmed by photographic observations.

Meteor echoes of FM-radio waves are observed and recorded automatically by microcomputers. The great shower of Draconids in October, 1985 was also recorder by FM-radio echoes as well as by visual observations.

4. Novae

At present, some Japanese amateurs are searching for galactic novae photographically, and extragalactic supernovae visually and/or photographically. In 1936, Kasuaki Gomi discovered a nova visually. He is one of the veteran observers of variable stars in Japan. Okabayashi also discovered another nova in 1936.

About 20 years ago, Honda began to take patrol photographs to search for novae. Honda and Yoshiyuki Kuwano have discovered a lot of novae. Matsuo Sugano also discovered one comet, one peculiar variable star and one nova during the past four years.

We are sure that discoveries and observations of comets, minor planets, meteors, novae and supernovae by Japanese amateurs will continue to increase and thus contribute to astronomy.

Amateur Astronomers' Contribution to the HIPPARCHOS Programme

P. Bacchus
40 rue Haute, La Grande Paroisse, F-77130 Montereau, France

The HIPPARCHOS satellite (High Precision Parallax Collecting Satellite) is designed to determine position, proper motion and parallax for a large number of stars. The precision expected (0.002 arcsec), and the absence of systematic errors with position in the sky arise from the following characteristics of the system:

- Measurements of angular distances between stars a long way apart (58°), are by comparison with a very stable angular reference. This reference is formed by an optical block consisting of two plane mirrors, rigidly mounted, forming an angle of 29° between them, and sending two separate stellar fields into the same telescope;
- the absence of flexure (thanks to weightlessness) and of thermal deformation (non-expansion material and thermal control) ensure that the angular reference is very stable;
- operation outside the atmosphere allows the theoretical resolution to be reached; refraction, dispersion and atmospheric scintillation are avoided; diurnal and seasonal effects that interfere with ground-based measurements are non-existent.

The optical system is a Schmidt, with the corrector plate being replaced by figuring of the angular mirrors. The two star fields, $0.9° \times 9°$, are imaged on a grating consisting of 2688 slits at 1.2 arcsec intervals; rotation of the satellite causes the grating to sweep across the fields. A single star is isolated by an electronic diaphragm (dissector) for several seconds, the intensity of its light being modulated by the grating. This is analyzed by photon-counting at a sampling rate of 1200 Hz. This give the phase of the modulation, which is equivalent to a measurement of the star's position relative to the grating. This takes place for all the programmed stars in both fields.

The catalogue of stars to be observed (the Input Catalogue) is comprehensive down to about magnitude 8; beyond that down to about magnitude 12 objects to be observed were drawn from suggestions made by experts in all fields of stellar astronomy. Two hundred requests for observations were made, totaling 800 000 stars, but there was a lot of redundancy, leaving only 231 776 different stars. From these the INCA catalogue of 114 880 stars was chosen.

Many of the stars to be observed (78%) are double or multiple. The various problems for observation and reduction that they cause are not discussed here. They also are difficult to identify, and it is here that amateurs come in.

The Index of Double Stars (IDS), which in theory lists them in a form convenient for visual observation, is not suitable for computerized treatment, for the following reasons:

- The positions are of low accuracy (± 0.1 m in RA, and $\pm 1'$ in Dec, and sometimes even more). This is not enough to set HIPPARCHOS or even an automatic transit instrument, and is not enough to find a star in a moderately rich field.
- Identifications are missing or incomplete in the major catalogues (DM, AGK3, SAO...). The IDS only gives the DM number, and sometimes omits that.
- In multiple systems, the meaning of pairs quoted (such as AB, BC and CD) is often ambiguous.
- The IDS has various inconsistencies, such as giving the number of one DM star and the position of another, or else quoting positions in incompatible ways, for epochs 1900 and 2000, for example.

J. Dommanget at the Belgian Royal Observatory therefore undertook the compilation of a new catalogue of double stars, CCDM (Catalogue of Components of Double and Multiple stars), arranged by component (one line per star). Initially, the subset of CCDM stars on the INCA list has been revised, but later work will extend to all double and multiple stars. In this way we hope to obtain a catalogue with precise positions, complete identification, good internal consistency, and unambiguous identification of components. In compiling CCDM, we have to resolve thousands of problems involving insufficient or doubtful identification. It is frequently sufficient to refer to various documents (general catalogues, surveys – particularly the Palomar Sky Survey or PSS – and original references). But it is often necessary to check identification with the sky, not by measuring it, but simply by checking what there is in a field a field a few minutes across. This applies to pairs that are closer than $5''$, which are difficult to identify as double on the PSS photographs.

The work of amateurs has proved to be very valuable in identifying several hundreds of systems on the sky. In West Germany, under Dr Bastian of the Rechen-Institut at Heidelberg, 20 amateurs at 8 observing stations have put in a lot of work, which, to date, has resulted in identifying 66 doubtful doubles. In doing this they have produce excellent finding charts. The exact positions of the doubles have been measured from the PSS.

In Belgium, the Royal Observatory has been helped by collaboration by several very active observers, particularly in resolving problems relating to forty-odd multiple systems incositently described in the IDS. G. Leonis, A. Jorissen and L. Louys made three trips to ESO in Chile between 1984 and 1986, and took more than 1000 plates with the GPO[1]. A fourth trip, by L. Louys and J.J. Doyle, is scheduled for next August. They measured and reduced their observations with either the ORB plate-measuring machine at Brussels or ESO's at Garching. In addition, the Belgian amateurs R. Barbier, J.J. Doyle and J. Bourgeois took part in two trips to La Palma

[1] The 0.4-m double astrograph, originally sited in South Africa – Eds.

in the Canaries, where, within the framework proposed by A.N. Argue at Cambridge, they made several hundred photometric measurements of poorly-known pairs, using the 1-metre telescope and a CCD camera.

In France, amateur activity has been guided by the Double Star Section of the S.A.F. and its Director, P. Durand will describe in detail the methods used and their results.

The S.A.F. Double Star Section's Contribution to the INCA-HIPPARCHOS Programme

P. Durand
Ecole Marcel Sembat, F-18000 Bourges,France

The Double Star Section of the Société Astronomique de France has been in existence since 1980. It has set up observing groups covering various areas: close binaries, photographic measurements, re-observation of neglected pairs, and astrolabe measurements. All this has been thanks to our sceintific advisors, professionals specializing in double stars: Muller, Couteau, Soulié, Bacchus, and Dommanget. The Section meets twice a year, and it was at Lille in 1984 that J. Dommanget suggested that we should collaborate in the programme of identifying doubtful doubles for the input catalogue for the future satellite Hipparcos.

The first observations were made after the meeting in November 1984, from Paris and with amateur equipment, thanks to an initial list prepared by Prof. Bacchus. After several months we could tell that our equipment was insufficient, and that photocopies of the Palomar Sky Survey were not accurate enough. We needed more powerful instruments and a different organisation to resolve the ambiguity in identification that can arise by two close stars being confused and given the same coordinates, by the presence of a cluster, or by the existence of a faint star that is not given in the catalogues. In all these cases, identifying the star means being able to give it a designation, allowing it to be found on an atlas, and obtaining better coordinates so that the ambiguity is removed. It is a question of astrometry.

In November 1985, the decision was taken to request three allocations of telescope time for the summer of 1986: at Nice (50-cm refractor), Observatoire de Haute-Provence (80-cm reflector), and Pic du Midi (1-m reflector). Eight to ten nights were allocated to us on each telecscope. The 4 observers were equipped with the AAVSO, Papadopoulos and PSS Atlases, and fifty-odd forms, which permitted the field to be identified on a PSS plate by x and y rectangular coordinates, gave the equatorial coordinates and details of the double required for identification purposes. A chronometer allowed the position to be related to reference stars by using the $\Delta\alpha$ and $\Delta\delta$ differences, from the known size of the instrument's field. On site, the teams planned their observations, arranging them in order of increasing RA, and distribtuion throughout the night, and familiarizing themselves with an observatory telescope that was new to them by establishing the fields of the various eyepieces.

Work at the eyepiece consisted of setting the telescope to the given coordinates, corrected for precession, checking the finder field against the AAVSO chart, centring the object, recognizing the main field with the help of the PSS, examining the double at a power of 350 to 600 in order to identify it, and making a sketch. The next day, in the office, the double was located on the PSS by using the observation, and new (x', y') coordinates were established. The completed forms were sent to Prof. Bacchus for the identification to be checked and for more precise coordinates to be calculated, possibly with reference to a catalogue.

Table 1. Results of three 1986 programmes

	Observations		Identifications		
Place	planned	carried out	confirmed	doubtful	Not seen
Nice	69	51	44	2	5
OHP	51	48	35	5	9
Pic	43	28	23	5	0

Not all the identifications apparent to observers were retained. We have found that what appeared to be a simple problem hid actual complexity on the sky. It is not just by chance that these doubles have been little observed and have remained poorly identified.

This amateur/professional collaboration has enabled us to develop and allowed us to contribute in the area where we are most at home: that of observation. It has also taught us to work more rigorously, and more methodically, thanks to advice from Prof. Bacchus. Finally, it has also revealed some neglected pairs that are in motion and which ought to be measured accurately. It has also been a step towards other programmes, in particular it has laid the foundation for the success of the 1987 campaign, three trips to Nice and the Pic du Midi, which we hope will give as good results as those we obtained in 1986.

Working with Amateur Astronomers

Stephen J. Edberg
International Halley Watch, Jet Propulsion Laboratory,
California Institute of Technology, 4800 Oak Grove Drive,
Pasadena, CA 911909-8099, U.S.A.

Introduction

Amateur astronomers have a vast store of talent and expertise in making astronomical observations. When carefully channelled, their enthusiasm coupled with these characteristics can lead to the acquisition of large volumes of high-quality astronomical data. The Amateur Observation Network of the International Halley Watch (IHW) was organized to encourage comet observations by amateurs, standardize techniques whenever possible, and then collect and archive these observations for use by the astronomical community in the near and distant future as part of the whole IHW archive. The lessons learned from this experience will be useful to organizers who plan observation campaigns involving amateur astronomers in the future.

Background

The IHW was first conceived by Louis Friedman of the Jet Propulsion Laboratory (JPL). From its conception, the IHW has included the observations of amateur astronomers in its campaign plans (Brandt *et al.*, 1980).

In its final form, the IHW consisted of eight professional disciplines defined primarily by their use of astronomical technologies:

1. Large-Scale Phenomena (tail behaviour and morphology)
2. Near-Nucleus Studies (coma morphology)
3. Spectroscopy and Spectrophotometry
4. Astrometry
5. Photometry and Polarimetry
6. Meteor Studies
7. Radio Studies
8. Infrared Photometry and Spectroscopy

Amateur astronomers have the potential to make contributions to the first six disciplines, though a handful of the more naive wanted to try "ham" radio bouncing off Halley's Comet and one wondered about doing infrared observations. Ultimately, data for use in these six areas were received, ranging from large numbers of photographs and visual magnitude estimates from many observers to a single set of photoelectric observations from one observer.

Organizing amateur observation was more easily accomplished on the basis of technique, rather than technology. Six areas of effort were identified:

1. Visual Observations, including magnitude estimates and drawings
2. Photography, from short to long focal lengths
3. Astrometry
4. Spectrophotography
5. Photoelectric Photometry
6. Meteor Observations, including hourly counts, photography and spectrophotography

To standardize amateurs' efforts in these areas the *International Halley Watch Amateur Observers' Manual For Scientific Comet Studies* (Edberg, 1983) was prepared and published. Besides being distributed through the U.S. Superintendent of Documents, two private publishers reprinted it and it was translated or adapted to Chinese, Finnish, French, German, Hungarian, Italian, Japanese, Polish and Spanish by interested parties outside the United States.

For each of the areas of amateur study a chapter was devoted to detailing the correct observing procedure to use, based on a literature search and consultation with experienced observers. Alternate methods were sometimes presented. Required calibration techniques were included. Sometimes data analysis methods were also presented so observers could analyze their own work.

Besides an IHW registration form, standard data report forms requesting many specifics were included in the manual. These were set up to be understood without reference to the manual, if necessary. Space was allocated for presentation of the data, method used for obtaining them, and standardization or calibration techniques used. In areas such as photographic processing, a standard practice simply could not be imposed so the report form asked for details of the observer's personal procedures.

To better keep registered observers as well as the whole community of amateur astronomers informed, the *IHW Amateur Observer's Bulletin* was published and distributed free, courtesy of the Planetary Society. Circular letters were also used occasionally to inform registered observers of important events, updates in report forms, or in data reporting procedures.

The IHW ultimately coordinated observing campaigns on Comet P/Crommelin (1984IV = 1983n), Comet P/Giacobini-Zinner (1985XIII = 1984e), and Comet P/Halley (1982i, no Roman numeral designation as yet). Crommelin served as a trial-run target, to test the IHW archiving procedures (see Sekanina and Aronsson, 1985). Comet Giacobini-Zinner (G-Z) was added to the IHW program when the International Sun-Earth Explorer No.3 (ISSE-3) spacecraft was diverted to G-Z and renamed the International Cometary Explorer (ICE).

Observations on the Observers

Over 1400 amateur astronomers registered with the Amateur Observation Network. In addition, well over 200 contributors sent in observations by no registration form (the majority of these were meteor observers). There were a total of about 1000 data contributors, ranging from beginner to very experienced.

In terms of observing technique it was, in a few cases, more difficult to get experienced observers to observe IHW requirements than novices. Many experienced amateur observers have difficulty in maintaining an unbiased, scientific attitude about their results or their methods of obtaining them. In the area of magnitude estimation there were difficulties in getting observers to agree on what standard comparison stars to use and that poor conditions of atmospheric transparency could affect their results. In a few cases, non-IHW reporting methods made their data difficult to enter in the IHW computer archive.

The amateur community, as a group, does not have a good understanding of the scientific method and the need for calibration. This was reflected, especially among photographers, by their failure to supply requested calibration photographs of selected astronomical targets. Perhaps too much was expected of contributors.

In order to include more than a tiny fraction of the submitted observations, the author has felt it necessary to relax many of the data reporting and calibration requirements and to prepare all the reasonable data for archiving. A strong warning will go with the archive, though, that data users must be selective about what they use and how they use it.

Observations on the Observations

The following table summarizes the data from amateur astronomers on the IHW's three campaign comets. The text following the table includes discussions of astrometry, photoelectric photometry, and meteor observations which are not listed in the table.

IHW Data Archives

Comet	Number of Observations			
	Visual Observations		Photographs	Spectra
	Mag. Ests.	Drawings		
P/Crommelin	197	0	0	0
P/Giacobini-Zinner	1000*	50*	50*	0
P/Halley	10000*	600*	2000*	52
	* estimated number			

The IHW data archives on G-Z and Halley will include a small number of copies on magnetic tape, a widely distributed collection of printed volumes, and a version on compact optical disk. The Crommelin archive was produced on magnetic tape and printed versions (Sekanina and Aronsson, 1985).

Visual Observations – Magnitude Estimates. The large number of Halley observations allow a number of interesting studies to be made. Edberg and Morris (1986) concluded that scatter in Halley's visual light-curve is most strongly influenced by observer experience and that experienced observers tend to see a larger coma and make correspondingly brighter magnitude estimates. Coma morphology also affects estimates but the instrument used has much less effect on estimates than these other factors. Inexperienced observers show a marked tendency towards estimate precisions of only 1/2 magnitude, whereas experienced observers are often precise to 1/10 magnitude. Much more investigation will be possible when the complete computer archive is available for analysis.

Visual Observations – Drawings. The drawings varied in quality from simple sketches of a comet's position in a star field to careful renderings of fine detail in the coma. Selected drawings may be included in the printed archive for comparison with professional images there. All suitable drawings will be catalogued in the archives. Little detailed study has been done as yet on the drawings.

Photography. Photographs, especially in the case of Halley, were made with focal lengths ranging from a few centimeters to over 3.5 meters. Both black & white and color emulsions were used. All photographs showing useful detail will be listed in the archives and amateur photographs filling gaps in professional coverage will be archived with the professional disciplines.

Astrometry. At the initiation of the IHW a small number of amateur astronomers were already making astrometric observations of comets and had established their reputations for high-quality measurements well enough to work directly with the professional discipline specialists. Newcomers to the field submitted their work to the Amateur Observation Network, which in turn checked those results with the professional discipline specialists. Regrettably and not surprisingly the data from novices were not usable. The residuals of their positions were returned to these observers along with a note of encouragement that their work would improve with practice.

Spectrophotography. Four observers submitted 52 Halley spectra. These low-dispersion spectra, obtained with a variety of instruments, will be listed in the archive.

Photoelectric Photometry. The very few of these observations made by amateur astronomers will be forwarded to the professional Photometry and Polarimetry discipline network for their ultimate disposition.

Meteor Studies. Hundreds or even thousands of hours of meteor counts have been made by over 300 observers. Often these were made in only one or two hour stretches, while a longer interval is more desirable. Two meteor photographs have been received. All these data have been forwarded to the professional Meteor Studies discipline specialist team for analysis and disposition to the archive.

Conclusion

Amateur astronomers can make significant contributions to observational campaigns using several methods of study. The organizers of such campaigns must understand the motivations and limitations of their amateur contributors and design the program accordingly. Early in the planning process, leading amateur observers in the campaign topic should be consulted on how best to organize the amateur community and on the methods of observation used by amateurs. Those contacts should be maintained throughout the campaign. Detailed instructions should be issued to the observers, but the organizer must be prepared for the unexpected.

Acknowledgements

Hundreds of amateur astronomers have contributed to make the IHW Amateur Observation Network a success. Mikael Aronsson, Carlo de Antonio, Mary Firth, and Charles Morris have all assisted in the data preparation or analysis. The research described in this paper was carried out by the Jet Propulsion Laboratory, California Institute of Technology, under contract with the National Aeronautics and Space Administration.

References

Brandt, J.C., Friedman, L.D., Newburn, R.L., and Yeomans, D.K., eds., 1980, *The International Halley Watch – Report of the Science Working Group*, Pasadena, NASA-JPL Publication TM 82181 (400–88)

Edberg, S.J., 1983, *International Halley Watch Amateur Observers' Manual For Scientific Comet Studies*, Pasadena, NASA-JPL Publication 83–16

Edberg, S.J. and Morris, C.S., 1986, "Observational Factors Affecting Studies of Comet P/Halley's Visual Light Curve" in *Proc. ESLAB Symposium on the Exploration of Halley's Comet*, Noordwijk, The Netherlands, ESA-ESTEC Publication SP-250, p. 609

Sekanina, Z. and Aronsson, M., 1985, *Archive Observations of Periodic Comet Crommelin*, Pasadena, NASA-JPL Publication 86–2

Amateur Astronomers and the Hubble Space Telescope

Stephen J. Edberg
HST Amateur Astronomers Working Group, 163 Starlight Crest Dr.,
La Cañada, CA 91011, U.S.A.

Background

In December 1985 the leaders of major amateur astronomy organizations in the U.S. met with the director and staff of the Space Telescope Science Institute (STScI) at their invitation to discuss the possibility of organizing a program for U.S. citizens who don't work professionally in astronomy to make use of the Hubble Space Telescope. Director Riccardo Giaconni's previous successful experience in cooperating with the American Association of Variable Star Observers (AAVSO) on studies of various objects with the orbiting High Energy Astronomy Observatory prompted him to bring together representative of the AAVSO, Association of Lunar and Planetary Observers (ALPO), Astronomical League (AL), International Amateur-Professional Photoelectric Photometry (IAPPP), Independent Space Research Group (ISRG), International Occultation Timing Association (IOTA), and Western Amateur Astronomers (WAA) to work out an approach to making HST observing time available to amateur astronomers, educators, and anyone else who might have a good research idea. Dr Giaconni offered to make some of his director's discretionary time available to qualified non-professional astronomers if the representatives present would design and carry out a program to find the good research ideas and have them ready for his selection. They would then be inserted in the HST observing program once the first six months of Guaranteed Time Observations had been carried out and the regular General Observing program had begun. With this as its charge, the representatives organized themselves into the Amateur Astronomers Working Group and have prepared a procedure for the selection of non-professional researchers to use the Hubble Space Telescope.

The Program

Certain restrictions were imposed on the AAWG by the National Aeronautics and Space Administration (NASA) and the STScI at the outset:

1. The proposed study must make use of the HST's unique capabilities, and must be impossible to do from the ground. These capabilities include its high resolution, extended wavelength range, and high sensitivity.

2. The proposer must not be a professional astronomer: the AAWG defined a professional as one with an advanced degree in astronomy or astrophysics or one who draws salary doing or assisting in research or data processing in these fields.
3. The principal proposer must be a U.S. citizen though co-investigators need not be.

The AAWG has designed a multi-step process to send proposals to Director Giaconni. First, an initial 250-word essay describing the proposed research is submitted to the amateur organization most familiar with that area of research. Proposal forms and a handbook (Littmann, 1986) describing the HST were distributed for this purpose. Proposers passing this initial screening by the organization are then asked to do a more formal, detailed proposal and justification, which includes some of the standard STScI proposal forms distributed to the professional astronomical community. Each organization will again screen these proposals and bring the best to a meeting of the AAWG in which all the proposals at the meeting will be prioritized. A formal STScI observing request will be prepared for each and then the list will be submitted to the Director for his selection of one or more for scheduling on the HST. When the observations are to be made the successful proposer will be invited to STScI to monitor the acquisition of the data and perform analysis on them. Ultimately the results are to be published in an appropriate journal and a report submitted to the sponsoring amateur organization and to the STScI.

It is the hope of the AAWG that a successful program on the Hubble Space Telescope will open the door to similar programs at major ground-based observatories.

References

Beaty, J. Kelly, 1985 May, "HST: Astronomy's Greatest Gambit", *Sky and Telescope*

Littmann, Mark, 1986 October, *Hubble Space Telescope Handbook for Amateur Astronomers*, Space Telescope Science Institute, Baltimore

Longair, Malcolm, 1985 April, "The Scientific Challenge of the Space Telescope", *Sky and Telescope*

Watson, Gary, 1986 January, "Building the Space Telescope's Optical System", *Astronomy*

The T.60 Operation at Pic du Midi

Christian Buil and Olivier Las Vergnas
28 rue Bonnat, F-31400 Toulouse, France

Since Spring 1983, the Pic du Midi and Toulouse Observatory has put the Gentilly 60-cm telescope at the disposal of amateurs. There was an initial experimental trial in 1982, and an amateur Programme Committee was set up in 1983. The "T.60 Association" took over the whole of the operation in 1984. So far, five hundred different amateur astronomers have used the equipment. The prime advantage of the T.60 is its site at the Pic du Midi, 2877 m altitude, one of the best observing sites in the world.

By giving new facilities to amateurs, the T.60 offers considerable scientific and educational rewards. Work that has already been carried out includes: determination of H-R diagrams of faint clusters, spectra of planetary nebulae, polarimetric and interferometric studies of nebulae, photoelectric and visual studies of variable stars, research into the chemical composition of faint stars, and observations of mutual phenomena of Jupiter's satellites.

The "T.60 Association" unites both users and other persons involved in the operation, and is responsible for all links with the Observatory. It is concerned with improving and providing accessory equipment for the telescope. It includes about 200 individual members, and a number of "associate members" who agree with the aims of the organization. A number of national societies, such as the S.A.F., the Association Française d'Astronomie and the Sociététe d'Astronomie Populaire come in this category, as do certain regional astronomical associations.

Amateurs use the telescope in teams of four, living at the summit alongside the professional astronomers, with whom there are frequently useful exchanges of ideas. The Programme Committee allocates telescope time, its main criteria being the suitability of the telescope for the proposed observations and that the proposed use is realistic. It seeks to make the telescope available to a maximum number of users.

Maintenance and accessory equipment are provided by a group of volunteers, the Technical Support Group. They are currently designing and commissioning a photographic plate-holder, micro-processor controlled setting, and a stellar spectrograph.

Annual "T.60 Colloquia" have been held since 1984, bringing together amateur and professional astronomers, with the aims of presenting results obtained and considering improvements and future developments. A "T.60 Letter" provides contact between the meetings. In order to prevent usage of the telescope from being confined to a small group of privileged people, from 1988, the Association will invite any inexperienced amateurs to take part in special training sessions.

Considerable amateur-professional cooperation has developed, so that numerous collaborative programmes have taken place. This has also led to other equipment being placed at the disposal of amateurs, such as the 1-m telescope near the T.60, which will now regularly be available to amateurs. Combined teams of amateurs and professionals will be able to use this instrument for such programmes as SONATE (analysis of seeing quality) and PG1115+08 (long-term surveillance of a triple quasar).

Improvements to the telescope are continually made. In 1988, for example, a high-performance CCD camera and an image-intensifier system are to be commissioned.

The network of observers using the T.60 continues to expand, now including those in other countries in both Europe and America. Spanish, Belgian, German, Italian and Canadian teams have already used the instrument. The Association envisages this form of international cooperation becoming even more widespread in future.

The Effects of Light Pollution on Amateur Astronomy

David Crawford
Kitt Peak National Observatory, P.O. Box 26732, Tucson, AZ 85726, U.S.A.

Abstract

The rapid increase in urban lighting, with the resultant increase in sky glow due to this lighting, has already severely affected both professional and amateur astronomy. It is not possible to observe as faint as was possible in the past from sites anywhere near large cities. Even those sites near smaller cities are compromised.

There are other adverse effects of outdoor lighting. One in particular affects amateur astronomers (more than professionals, in fact). That is light trespass from a (usually) nearby source shining directly into the amateur's observatory. It ruins dark adaption and can even affect photographic or photoelectric observations.

Astronomers are pushing for communities and individuals to apply "solutions", which can help greatly. These are: use the correct amount of light for the lighting task, not overkill; use controls on the lighting, such as shielding and correct placement (direct the light downward), and time controls; use low pressure sodium light sources whenever possible (especially in the vicinity of professional observing sites).

Without the widespread application of such solutions, amateurs, professionals, and even the general public will likely lose forever their superb view of the universe, as seen from a prime dark sky site. These solutions also maximize the quality of any lighting installation, often at lower cost than bad lighting.

An Amateur CCD Camera

Christian Buil, Guylaine Prat, Eric Thouvenot
28, rue Bonnat, F-31400 Toulouse, France

A CCD camera was built in 1984 using a linear detector (Thomson TH7810A) with 1728 pixels. The simplicity of this sort of detector enabled us to gain experience with CCDs. With a linear detector it is possible to carry out spectroscopy or even two-dimensional imaging by shifting the observed field at right angles to the photosensitive strip and by acquiring data in a regular manner (the scanning technique). The limitations of this equipment soon became apparent when trying to image deep-sky objects, so a second camera was built in 1985 using a CCD array, a Thomson TH7851 with 208 × 144 pixels. Numerous improvements were later made and the most advanced model is described.

The CCD Camera

Although the camera's principles are simple, it has very high performance, and technical solutions have made it very flexible in its applications, rapidly operational, and reliable. This has been largely because all the CCD's operations are controlled by a microcomputer, including charge transfer within the CCD. Windowing and binning, for example, are easily controlled by modification of instructions in a suitable program. Full computer-control has resulted in keeping the electronics to a minimum, with a consequent reduction in cost and ease of duplication. The computer is an IBM AT compatible, and the suite of programs is written in compiled BASIC. The greatest problems arise over the video output. Standard EGA graphics are used, allowing images of 640 × 350 pixels in 16 colours to be shown. VGA standards (320 × 200 pixels but with 256 colours) are more pleasant to use.

The CCD is a Thomson TH7552A, 208 × 144 pixels, each 30 microns square, with raster-type charge transfer. The detector can be operated to prevent overloading, thus allowing observation of regions close to bright stars. This array costs about 2000 Fr [approximately $340 or £200 – Eds.]. Although larger arrays can be used, problems may occur with the amounts of data that have to be stored and manipulated (12-bit digitization of a 512 × 512 array would require 512 k of memory).

Cooling is essential to minimize parasitic charges (dark current) – this detector will saturate in four seconds at 20°C – so the detector is cooled to –50°C by 2 Peltier-effect thermoelectric stages, using a closed water-cooling circuit and a heat exchanger. The array is enclosed in a chamber held at about 10^{-3} Torr to avoid condensation on the CCD's optical window. Practical integration times range from a few tens of seconds to thirty minutes.

The camera's read-out noise is less than 100 electrons. This is obtained by two-stage numerical sampling (reference array – video array), a unique technique that has proved to be highly efficient and also allows the electronics to be further simplified.

The computer generates all the CCD control signals, and stores the 12-bit digitized signal. The conversion time for one pixel is 25 microseconds. A "Quick-Look" can be displayed on the EGA screen immediately after scanning. The acquisition program includes all the usual image-manipulation and presentation tools (various convolutions, arithmetical operations, isophotes, profiles, false colours...). This allows the quality of the images to be evaluated immediately after acquisition. An arithmetic co-processor is used (Intel 80287) and optimization of the compiled code has allowed high performance to be secured in implementation of two-dimensional Fast Fourier Transforms, for example.

Initially images are stored on a 20 M hard disk and later backed-up onto 1.2 M floppies. One image occupies approximately 60 k. Read and write times to the hard disk are about 1 second.

Some Results

The camera has been used on various telescopes, ranging from a small 150-mm to the T 60 and the 1-m at the Pic du Midi. The objects observed obviously varied with different telescopes, but on every occasion the CCD produced some surprises.

In planetary observation, the high sensitivity of CCDs allows short exposures, partly freezing atmospheric turbulence. The large dynamic range allows low-contrast details to be studied. This is one of the reasons we use 12-bit digitization of the video signal from the CCD, giving 4096 grey-levels. Such a range is unknown in photography. A standard CCD has a spectral range from 4000 to 10 000 Å, and it is very instructive to observe the planets over such a range. Jupiter, for example, shows completely different appearances in the blue (high-contrast belts, few details) or in the infrared (low-contrast, but a mass of detail right up to the poles). The same planet observed with the 1-m telescope through a filter with a narrow pass-band close to 8900 Å, corresponding to a methane absorption, shows a completely different aspect to the normal visual image: the poles are practically the brightest areas, while the temperate zones are very pale, and spots like the GRS appear that have no counterpart at other colours. We have also obtained a tri-colour image of Jupiter. The procedure is simple: one only has to acquire images of the planet at short time-intervals through three different filters (in our case: blue, green and infrared), and then numerically superimpose these images associating each with one of the primary colours. The result is a highly arresting colour image of Jupiter's cloud patterns. The radiometric quality of CCD images allows very faithful colours to be obtained after suitable calibration, and this would certainly not apply to ordinary colour film. We hope to obtain the first tri-colour images of galaxies in the near future.

The most satisfying images were those obtained with a 150-mm, f/5 reflector, typical of instruments in amateur hands. The Ring Nebula (M57) is recognizable after only 5 seconds' integration. After 10 seconds the central star is visible; a 10-minute integration shows the ring more or less saturated, but the very much weaker, outer ring is detectable. The small, 15.2-magnitude galaxy IC 1296, 4 arcminutes from M57 is easily visible. Comparing this image with the Palomar Sky Survey, stars of visual magnitude 20 are visible. Not bad for a 150-mm telescope!

This must be qualified, however. The images were acquired without a filter, but for work of any astrophysical significance the CCD's range must be limited by BVRI-type filters (with pass-bands of about 1000 Å). With such a filter, and a 10-minute exposure on the same instrument the limiting magnitude is about 17 in the red. A 400-mm telescope ought to be able to reach magnitude 23 with a one-hour exposure, which is practical with our present CCD cooling arrangement. At this level of detection it would be possible to study variable stars in M31 (which is well-resolved in just a 10-minute exposure on the 150-mm telescope).

A composite of two images, each of 10-minute integration, of the pair of galaxies NGC 5194 and 5195, better known as M51, clearly shows the material extending towards the west from NGC 5195. When the image processing is "pushed", these extensions can be traced out to a surface magnitude of 25 per square arcsecond (according to the photometric study carried out by M.S. Burkhead: Astroph. J. Sup., **28**, 147–184, 1978). Such a magnitude is obtained because each pixel integrates energy arriving from a angle of more than 8 arcseconds on the sky. It should be noted that we obtained this image under considerable moonlight and quite a hazy sky. Under better conditions and with suitable correction, it would appear possible to reach a surface brightness of magnitude 27. This result also shows that it is realistic to observe faint objects with considerable sky background.

Conclusion

As we have seen, the performance of CCDs is high and very appealing, out-classing photography in most applications, except for wide fields. Few amateurs are able to exploit photography to its limits. Work by David Malin, for example, has shown that image-amplification techniques allow considerable gains in detection to be achieved. But these gains are only acquired after a long series of careful trials and more hours spent in the darkroom than out observing. CCD images, on the other hand, all give optimum results. One importance aspect is that a CCD image is easier to exploit than a photograph, which too often is simply regarded as a work of art. A CCD image will give more information and will result in greater satisfaction at the end of the day. In the not-too-distant future it is even possible that an "image bank" will be set up, that one can consult by obtaining diskettes. Amateurs will probably have common methods of image processing, which will facilitate comparison of results.

It is clear that the introduction of the CCD is as important an event for astronomical observation as the introduction of photography was at the end of the last century.

A Spectrograph for the T.60 at Pic du Midi

Daniel Bardin and *Didier Buty*
Villa Saint-Clair, Chemin de la Marlene, La Pomme,
F-13011 Marseille, France

For several years now the 60-cm telescope at Pic du Midi has been made available to amateur astronomers. We felt that this telescope, which fulfils its role of gatherering light very well, could be fitted with equipment to analyze light. To this end the members of the Teachers and Astronomers Liaison Committee (which includes one of the authors of this communication) collaborated in producing and perfecting a spectrograph suitable for the T.60.

Of basic construction, this equipment enables spectra to be recorded on photographic cassettes. It should be transported to the summit at the end of 1987, and will then be available for any amateur projects that may want to use it.

With its focal length of 2100 mm, the T.60 therefore has a focal ratio of 3.5. In imaging the primary image onto the slit, it needed a collimating lens of f/3.5, with an exit pupil appropriate to the area of the grating. In the same way, the field lens of the camera objective needed to receive all the light that had passed through the system.

The components of the spectrograph are as follows, where the numbers refer to the Figure.

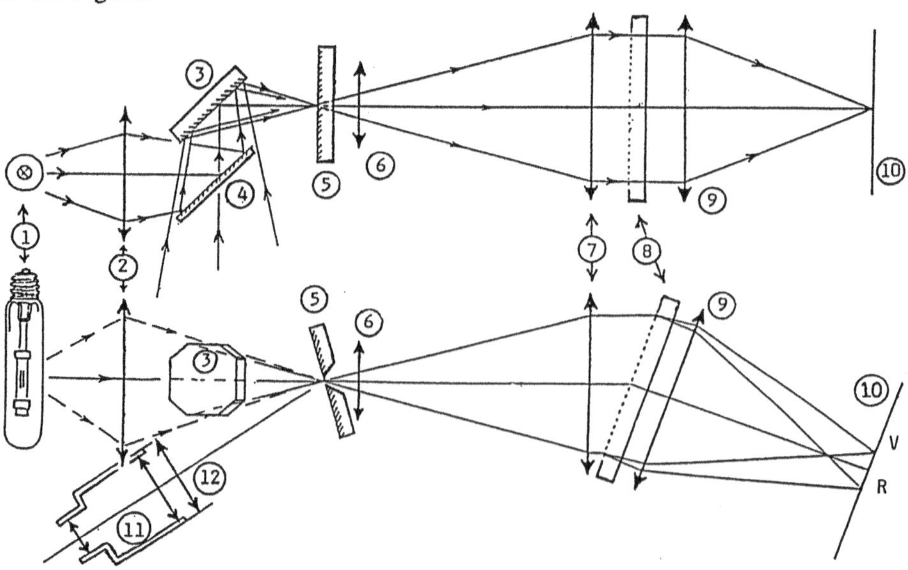

1) argon lamp (reference source); 2) lens focussing light from the argon lamp onto the slit and ensuring that the collimator is fully covered; 3) tilting mirror that enables the slit to be swept (in RA), and thus broadens the spectra; 4) mirror that is only used when the reference spectrum is taken; 5) reflecting slit, the reflecting surfaces of which return an image of the field to the small guiding telescope; 6) field lens; 7) collimator; 8) diffraction grating; 9) camera lens; 10) film; 11) centring and guiding telescope; 12) collimator for this telescope.

As far as cameras are concerned, at present these are standard bodies and lenses brought up for each individual amateur project. For example, a lens of 200 mm aperture, f/3.5, is a good choice and covers a spectral range of 275 nanometres on a 24 × 36 mm frame. A lens of this type was used in the commissioning tests and allows a resolution better than 0.1 nm in wavelength.

Mounted on the T.60 and in the form it has now, the equipment has already been used (and will be used in future) for various projects, among which we may mention:

1) tests of spectral resolution and sensitometry; comparative tests of gratings, black-and-white and colour emulsions
2) classification of spectra
3) spectral shifts and radial velocities
4) spectral variability: variable stars, emission lines, reinforcement or weakening of lines, abundance problems, stars with high energy losses, spectroscopic doubles, etc.

Undoubtedly this spectrograph is not perfect, it could even have been designed differently. Being suitable for the T.60, however, it should allow amateurs to undertake work in the field of spectroscopy that is comparatively unexplored by them as yet, and at a remarkable site: the Pic du Midi.

The author would like to thank all his colleagues and friends who have accompanied him on various trips, and whose assistance has enabled the project to brought to a successful conclusion. Our thanks are also due to the members of the T.60 Association, to the allocations committee and to all the staff at the Observatory.

Using an Image-Intensifier in Astronomy

Thierry Midavaine
Appt. 1199, 142 Bd. Masséna, F-75013 Paris, France

Abstract

The use of an RTC XX 1390 second-generation intensifier (which consists of a S20R photocathode, a micro-channel plate and a screen) was described. In visual work, very faint objects become visible. Photographically, an 18th-magnitude object has been detected through a layer of cirrus. The effective focal length used was 20 m, giving a resolution of 1 arcsecond, thanks to the short exposure times needed. With a Celestron C8 telescope, the coma of Comet Halley was photographed with 2-second exposures, allowing transitory phenomena to be detected. Used in conjunction with video techniques, high time-resolution images could be obtained, opening up the possiblity of new forms of observation of faint objects.

References

1. Caron, R., Koutchmy, S., Sarazin, M.: "A study of disconnection velocities in the plasma tail of P/Halley", paper presented at this Colloquium
2. Maley, P.D.: "Achievements in astronomical video data collection", paper presented at this Colloquium
3. Dahlmark, L., "Searching for variables, and constructing a slit-less spectroscope", paper presented at this Colloquium
4. Midavaine, T.: in *Astronomie: le guide de l'observateur*, ed. Martinez, P., Vol. 2, Chap. XIX, Paris, 1987
5. Mauro, J.A., ed.: *Optical Engineering Handbook*, General Electric Co., Syracuse, N.Y., 1966
6. Midavaine, T.: *Observations et Travaux*, in press

Amateur Polarimetric Research

Dietmar Böhme
DDR-4851 Nessa 11, PSF 93, GDR

Abstract

The construction of both single- and twin-channel polarimeters was described. Refractors are suitable for this form of work, and Cassegrains may be used, but Newtonians are not recommended, because of the polarization it introduces. A summary of objects and their degree of polarization was given. Examples of variation in the amount of polarization and its correlation with visual observations of μ Cephei were given.

References

A full account (in German) has been published in *Sterne und Weltraum*, **25**, p.544–9, 1986
Observations of μ Cep are given in *Die Sterne*, **60**, p.315–8, 1984

The following 5 papers, the texts of which have not been received, or have already been published, were also presented

Searching for Variables, and Constructing a Slit-less Spectroscope

Lennart Dahlmark
Montlaux, F-04230 St. Etienne les Orgues, France

Methods of searching for variables using a short-focal-length (30 cm) and stereo and blink comparators were described, together with the design of a simple slit-less spectrograph suitable for a Newtonian or Cassegrain telescope.

Solar Spectroscopy with a CCD

Olivier Saint-Pé
Observatorie de Meudon, 5 Place Jules-Janssen,
F-92195 Meudon Principal, France

The design of a spectrograph having spatial resolution and using a linear CCD was described, together with the optical solution adopted (an Ebert device) and the processing applied to the data obtained.

Reference

Saint-Pé, O., Collot, L., Buil, C.: "Spectroscopie à CCD", *Obs. et Travaux (S.A.F.)*, (8) 17–18, 1986

Obtaining Extraterrestrial Dust by Balloon Probes

L. Baufette
Association Astronomique de Franche-Comté, Parc de l'Observatoire,
34, av. de l'Observatoire, F-25000 Besançon, France

The "Icare" experiment, involving a balloon at a height of about 30 km was described. The gondola was made of polystyrene, the reasoning being that micrometeorites, by virtue of their velocity, would embed themselves in this to a sufficient degree that they would not be lost during the payload's recovery and subsequent handling.

Electron microscope examination revealed dust particles collected during the flights, and micro-analysis suggests that these are truly of extraterrestrial origin.

Making a Telescope Mirror

J.M. Becker
20 rue du Petit Chinois, F-25200 Montbélliard, France

A video presentation of all stages in the manufacture and testing of a 200-mm mirror. [The video cassette may be obtained from ASTAM, Viry, 39360 Vaux-les-Sainte Claude, France].

Searching for Comets

Ben Mayer
1940 Cotner Ave., Los Angeles, CA 90025, U.S.A.

Methods of searching for comets, novae and other events using projection and video blink methods, and simple orientation aids were described. The methods are covered in the following references:

1. Liller, W., Mayer, B.: *The Cambridge Astronomy Guide*, Cambridge University Press, 1985
2. Mayer, B.: International Halley Seminar, Alden Biessen, Belgium, 1986 Oct.25
3. Mayer, B.: 20th ESLAB Symposium, "Exploration of Halley's Comet", Heidelberg, 1986 Oct.27–31
4. Mayer, B.: *Mercury*, (Journal of the Astronomical Society of the Pacific), 1987 July/August

Achievements in Astronomical Video Data Collection

Paul D. Maley
International Occultation Timing Association, 15807 Brookvilla,
Houston, TX 77059, U.S.A.

Observation by amateurs has been traditionally viewed with some suspicion by the professional astronomical community, especially in the case of transient events(1). For example, data from variable star monitoring where information from many observers can be integrated over long intervals, have a higher weight than visual observations of an asteroid occultation. Serious amateurs who choose to contribute to the science face a continuing challenge to maintain the highest standards.

In an effort to upgrade the quality of information transmitted to professional institutions, we have experimented since 1984 with a baseline design of mobile video equipment which can be coupled to objective lenses ranging from 28mm to 625mm aperture (2).

A composite baseline system consists of the objective which passes light through to the intensifier, which then continues to a relay lens that transfers the image into a television camera. Once in video format the signal is output to a monitor and video cassette recorder. The system we use now consists of a RCA Ultricon camera, Litton M841 Pocketscope image intensifier, a relay lens, and Celestron 8 telescope. The objective has been interchanged with 100mm length lenses depending on the object under analysis.

We find that video achieves a level of accuracy better than the human eye but somewhat less than photometry. Time resolution is .03 seconds, consistent with a 30 frame/second recording rate, while a character generating accessory can display the time to either .1 or .01 seconds. Video offers distinct advantages of reusable recording material, playback and reviewing, slow motion or rapid frame advancement. These are characteristics which cannot be provided by photometric methods. Seeing effects may be isolated by comparison with pixel elements in other parts of the field not under study, such as the moon's surface in the case of lunar occultations and other stars, as with minor planet occultations.

Power can be supplied from the mains (110V AC, 60cps) or from 12V DC creating the ability to transport the hardware for field useage. However in mid 1987 portable video cassette recorders are rapidly disappearing from the market place and are being replaced by the Camcorder models which at present are not adaptive to low light astronomical use. The Celestron 8 system permits stellar images to magnitude 10.5 to be recorded under the best sky conditions using an 18mm aperture intensifier tube. With a 25mm tube, sensing to +11 is possible, though the price may more than double for the increased size of that component. Performance is impacted if the tube is not new, contains blemishes, or is of an earlier generation. A fast "matched" relay lens is essential to maintain the signal level and field symmetry as the image is passed through the optical train(3). Intensification is not required on such targets as solar eclipses and occultations of stars brighter than 5th magnitude.

We have worked on the following aerospace and astronomical applications of video to the science in an effort to demonstrate the ability of amateurs to contribute useful data on a variety of subjects (4). Our first use of a intensifier in 1972 accidentally discovered a brief burst of activity from the Giacobinid radiant of meteors below naked eye threshhold (5).

ECLIPSES: high precision detailed imagery of Baily's Beads have been obtained during the annular eclipse of May 30, 1984 (Atlanta, Georgia USA) and from the total eclipse of November 23, 1984 (Kwikila, Papua New Guinea) using unintensified video and projection techniques(6). Direct imagery of a one-second total-annular eclipse was achieved in Gabon on March 29, 1987. These data are being employed in the U.S. Naval Observatory solar radius variation study project, in which IOTA is a major participant.

OCCULTATIONS: numerous grazing lunar occultations have been observed since 1984 including those of Alpha Virginis and Eta Tauri in 1987. Clear definition of instantaneous and slow disappearances and reappearances have been obtained thorough evaluation of the exact number of video frames required for each phenomenon. The video records also serve as key anchors in reducing the more numerous visual observations. Duplicity of close components have also been recognized in the case of stars whose separations are 0.1 arcsecond or greater. A video record of the grazing occultation of Alpha Libra from Sudan during the May 4, 1985 total lunar eclipse was completed with a similar system (100mm f/10 objective). Analysis is underway to determine if this event will produce the most accuracte ground based determination yet of the lunar polar diameter(7). As many as 5 video systems have been simultaneously used on a single graze event.

Because of the rarity of asteroid occultation events, several have been attempted by us in which any clear reduction in magnitude of 0.3 or more would have been detectable, but none have yet been seen. Other IOTA observers have recorded at least 2 minor planet occultations with a portable system of similar design (8, 8A).

COMETS: Stellar appulses to Comet P/Halley have been recorded on video on several occasions. While no occultation was seen, the lack of an outage allowed an upper limit to be set on the dust density in the vicinity of the nucleus prior to the Halley spacecraft fly-by. The comet was also video taped from the northern Chilean desert in April 1986, with a 12.5cm f/2 objective; and a stellar appulse to Comet Giacobinni-Zinner has also been recorded. However our optical search for new comets conducted between 1975-1987 has been unsuccessful. We have employed low light television only within the last 3 years to assist in acquisition. Although this search has not achieved discovery of a comet, we conclude that image tubes fail to integrate enough light to permit distinguishing diffuse comets from point sources in order for them to serve as comet locators.

AEROSPACE EVENTS: Man-made space targets have proven more frequent and productive in which to apply the baseline system. A recent analysis of the sunlight reflecting behavior of a number of Soviet spacecraft have aided in the proof needed to demonstrate that claims of a new astrophysical discovery (the so-called Aries or Perseus Flasher) are more likely to be glints from artificial earth satellites (9). Light curves have been constructed using video recordings that were instrumental in the determination of the optimum rendezvous attitudes required from which to snare the disabled Westar and Palapa spacecraft in 1984 by a Space Shuttle crew. The reflectivities were obtained under a variety of phase angles.

This video arrangement has been used to study the reflectivity of over 30 geostationary satellites (with a 31cm reflector), and to determine details on the stability and on-orbit behavior of Soviet surveillance craft using a 100mm objective (10). Our two-year analysis of 14-day Soviet reconaissance satellites discovered key visibility seasons from which to monitor separation of the main body into recoverable capsules and earth resources subsatellites (11). Also in 1985, the NOAA 8 spacecraft suffered an internal explosion. The true rotation rate masked within a complex pattern of light changes was clearly found from our video data, and from that an initial velocity imparted by the explosion.

Our video was the only amateur astronomical data source contributed to the Max Plank Institute magnetospheric experiment known as AMPTE in 1984. Imagery which has been frame stacked and enhanced by an offline special processor clearly showed a solar wind shock front developing around a barium release. Another outstanding application has been the NASA acceptance of our suggestion to perform a controlled waste water dump from the Space Shuttle at a terminator crossing under dark sky conditions. This created a spectacular naked eye artificial comet seen by thousands of ground observers; it unexpectedly proved that contamination from routine expulsion of waste products from manned space vehicles could prove to be detrimental to astronomical experiments conducted in real-time or a number of orbits later. A plume 28 km long and 1.5 km wide was shown to exist, which later actually recontacted the Shuttle! Our video has been incorporated in an official Detailed Test Objective in which NASA hopes to study future effluence behavior under a variety of attitudes (12).

Ours was the first video taken of a Space Shuttle reentry, and it verified a theoretical chemiluminescent effect, which proved that a blunt nosed body entering the atmosphere causes a residual glow in its wake (13). This is extremely important in understanding the behavior of the upper atmosphere. One unusual phenomenon created by the 1908 Tunguska encounter was the highly brightened night sky which persist for several days following the event. A discovery of primary and secondary fireballs attending the Space Shuttle at high Mach numbers was evidenced on our tape. The most dramatic application came on April 6, 1984 as we successfully monitored the reentry of the Shuttle external fuel tank from the slopes of the Mauna Kea volcano in Hawaii (14). A complete detailed evolution of each major fragment was clearly recorded and its trajectory determined.

The baseline video system can be operated by one person and most components may be hand-carried onto a commerical aircraft. It has been tried and proven under a variety of conditions. While video hardware is not consistent in terms of frequency and signal characteristics, manufacturers do supply VCR, camera and monitor components in SECAM and PAL-B. A video tape recorded on SECAM cannot play on a recorder/monitor which is designed for the American NTSC, so that the electronics must be considered with this limitation in mind. While tape formats can be interchanged, the recorded signals must be converted between systems; this can be an expensive drawback.

We encourage the proliferation of this type of technology into the hands of the international amateur astronomical community so that they may apply it to the science. The ultimate goals may be educational or research oriented but amateurs will find an increasing respect as their tools lift their status and ultimately their treatment as co-investigators from a heretofore lower case role.

Sources in Europe for image intensifier assemblies and relay lenses are from Litton Precision Productions International, Inc. Inquiries may be made to this company through the following offices:

58 Rue Pottier F 78150 Le Chesnay FRANCE	(Contact Bernard Desprez Telephone: 33-1 395 52104)
Oberfohringer Strasse 8 D-8000 Munich 80 FRG	(Contact Volkmar Koerner Telephone: 49-89 980547)
6 First Avenue Globe Park, Marlowe Bucks, SL7 1YA UK	(Contact Bob Nichols Telephone: 44-628 56060)

We would like to gratefully acknowledge Litton Electron Devices, Tempe, Arizona USA for the use of a prototype image intensifier for testing purposes, and to Peter Manly for hardware design assistance and review of this paper.

REFERENCES

1. H. Reitsema, SCIENCE, 1979, Vol. 205, p. 185.
2. P. D. Maley, PULSAR, September-October 1985, No. 656, p.220.
3. P. Manly, "Coupling Intensifiers to Television Sensors" (submitted to the Eighth Annual IAPPP Symposium, February 1987).
4. P. D. Maley, SPACE EDUCATION, Vol. 1, No. 9, May 1985, p.415
5. B. Marsden, IAUC 2452, October 20, 1972.
6. P. D. Maley, L'ASTRONOMIA, September 1986, No. 58, p.54.
7. D. Dunham, OCCULTATION NEWSLETTER, Vol. III 1985, p.246.
8. D. Dunham, OCCULTATION NEWSLETTER, Vol. III 1984, p. 182.

8A. D. Dunham, OCCULTATION NEWSLETTER, Vol. III 1984, p. 102.
9. P. D. Maley, ASTROPHYSICAL JOURNAL LETTERS, Vol. 317, L39-L44, 1987.
10. N.L.Johnson, THE SOVIET YEAR IN SPACE 1984, 1985, p.33.
11. P. D. Maley, PROCEEDINGS OF THE 1985 SYMPOSIUM ON SPACE C3, USAF Academy, Colorado Springs, May 12, 1985, p.133.
12. P. D. Maley and L. Leger, NASA FLIGHT TEST AND SUPPLEMENTAL OBJECTIVES DOCUMENT, DTO #0330., 1986.
13. P. D. Maley, ACTA ASTRONAUTICA, Vol. 12, No. 10, 1985, p.755.
14. P. D. Maley, SPACEFLIGHT, December 1984, Vol. 26, No. 12, p.446.

Video Grazing Occultations

David W. Dunham
P.O. Box 7488, Silver Spring, MD 20907, U.S.A.

Grazing occultations are always visible at the projections onto the surface of the Earth of the lines tangent to the lunar limb that are created by the motion of the Moon relative to the star being occulted.

Most information about the lunar profile continues to be provided by visual observers using simple equipment. Everyone is to be encouraged to join expeditions to observe grazes, because the accuracy of the observed profile is proportional to the number of stations. The use of grazes improves knowledge of the lunar profile (needed for the analysis of solar eclipse timings), data about close binaries, and galactic rotation from stellar reference-frame determination. Contact timings made near the edges of a solar eclipse track, for example, enable the diameter of the Sun to be derived relative to the lunar diameter, the latter being determined from occultations.

Using video equipment, it is possible to record such an event with an accuracy of ± 0.04 s (single frame resolution) and at the bright limb, whereas a visual observer will have problems recovering the star at a bright reappearance, and moreover will only attain ± 0.1 s time-resolution. Observing with a 20-cm telescope and a "standard" – commercial – colour video-camera, it is possible to record stars down to magnitude 8.

Four grazes recorded on video were shown:

1: α Librae, 1985 May 4. This graze was used to observe at both limits to determine an accurate polar diameter of the Moon.

2: δ Cancri, 1981 May 10. The first multiple event ever recorded on video (by A. Fiala), while 18 other stations observed visually. Misses occurred at the 2 northernmost stations, but multiple events at nearly all of the others. This event was important for determining the lunar profile as the latitude libration was similar to that found at solar eclipses.

3: σ Scorpii, 1986 March 30. This event demonstrated the possibility of resolving double stars with separations as small as 0.01 arcsec.

4: α Scorpii, 1986 July 18. Observed entirely on the bright limb and one of the best-observed events to date, with about 170 timings from 20 stations for 6 separate teams in California, and 15 stations near Miami, Florida. Five tapes of the graze were obtained. Steve Edberg, observing near San Diego, recorded 8 contacts. Dark-limb grazes are even better, when relatively long dimmings, partial blinks, and faint flashes, caused by the large angular diameter of Antares, 0.04 arcsec, are most apparent. The only stars occulted by the Moon with larger angular diameters are the Mira variable R Leonis (0.067 arcsec) and the Sun. Events involving Antares' greenish fifth-magnitude companion can be seen during dark-limb southern-limit grazes at night.

An Improved Method of Photographic Measures of Double Stars

D. Rodriguez[1], *J.A. Bernedo*[1,2] and *D. Valls-Gabaud*[1,3]
[1] Agrupación Astronómica de Madrid, Apartado 1039, E-28080 Madrid, Spain
[2] Planetario de Madrid, Parque Tierno Galván, E-28045 Madrid, Spain
[3] Institut d'Astrophysique de Paris, 98 bis, Bld. Arago, F-75014 Paris, France

Summary. We present a new technique in double star photography leading to high accuracy measurements in amateur-size instruments. The trailing method has been improved by the use of a chopper that cuts the trail into a number of segments. The chopper period is accurately known, so it is the length of each segment on the plate. In this way the scale factor is *measured* in *each* exposure. Fine-grain films, such as hyper-sensitized Kodak 2415, are used and all measurements are performed at the microscope. The same technique may also be applied to CCD cameras. A statistical analysis of errors shows that $\Delta\theta \leq 30'$ and $\Delta\rho/\rho \approx 0.01$ for most binaries with $\rho \geq 3''$. This method has been applied to professional-amateur joint programs dealing with multiple systems with variable components and a survey of poorly-observed wide binaries.

1. Introduction

We started this work to measure the parameters of a series of wide double stars of which the most recent data found in catalogues were obtained in years around 1910, and remained practically unobserved since then. The second project was to detect possible changes in systems with variable components . From the amateur's point of view, the means to carry out this work are modest, even with the newest techniques [*Ref. 3*], so it was fundamental to design a new method to insure the maximum accuracy possible, and above all, a reliable way of accounting for the errors in each stage of the measuring process. This point is particularly important for two reasons : (i) In the computation of orbital elements, the *rôle* played by the uncertainties in each measure is crucial [*Refs. 2,11*], (ii) The knowledge of errors could help to understand some trends that frequently appear in photographic determinations of P.A. and separation (*e.g.* the Kostinsky effect [*Ref. 5*], and the dependence of the P.A. error on separation [*Refs. 9,12*]).

A second part of this work is the study of the same stars with small professional instruments (Observatoire de Paris–Meudon), in order to establish a comparison between results and check for possible systematic errors and to further improve the whole process. The report of this part will be described elsewhere.

2. The Method

2.1 *Photographic recording.* The use of visual observing systems were discarded due to the high cost of the equipment involved for the desired accuracy, or to the high intrinsic errors expected. For instance, new filar amateur micrometers [*Ref. 10*] provide $\Delta\theta \approx 20'$ and $\Delta\rho \approx 0.''5$, while the double image micrometer [*Ref. 3*] gives slightly better results but is limited to bright pairs. Moreover, in most micrometer techniques, but the diffraction grating one [*Refs. 3,6*], the calibration remains difficult if high accuracy is desired.

We used a film hyper-sensitizing technique, developped at the Agrupación Astronómica de Madrid for 4 years and which has shown a very high performance. We treated the micrographic film Kodak 2415 with pure hydrogen hyper-sensitizing process at a pressure of 2 $kg\ cm^3$ and kept at 36° C during 48 hours, resulting in a very high resolution, 1000 ASA sensitivity film.

On this film the following must be recorded in order to use our technique: (i) The images of both components (with the proper telescope tracking), (ii) A trail (without tracking), to define the East-West direction (essential for position angle measurements), and (iii) The scale factor : a measurement with which to convert all linear dimensions into angular quantities.

2.2 *Measure of time.* The scale factor is the key point of this study. Most observers take "fixed" pairs or the Pleyades to calibrate their measurements [*Refs. 1,3,4,6*]. This technique is right only if there is a *complete* sample of *well-measured*, fixed pairs for all separations *and* magnitudes. It can be useful to take some standards from high-quality measures [*i.e. Ref. 7*], but there are very few of these at magnitudes fainter than 8.

Taking into account that it will affect *all* measurements, its accurate determination is essential. We tried first electromechanic shutters (EMS) driven by electronic clocks, but their dispersion were too high for our purposes. We finally came out to a solution which avoided these problems : *a rotating shutter driven by a synchronous motor.* Quite differently from others methods, we place the shutter in front of the telescope aperture in a way that the light path to the primary mirror/objetive is interrupted. This configuration has the disadvantage that not all the incoming light is cut at the same time, so the shutter dimension in the direction of motion has to be comparable with the telescope diameter. In this way we obtained images like the shown in *Figure 1* for Cor Caroli (α CVn). The extinction of the trail is not instantaneous, but occurs in a short interval as it is the case for the reappearance of the trail. To prevent this phenomenon from impairing the measurements, it was necessary to choose the right relationship between shutter dimensions and the distance of its rotation axis to the centre of the telescope aperture, as shown in *Figure 2.* We found that the best trail is obtained when the segment completely dissappears, leaving 1/10 to 1/5 of a segment before a new one (*Fig. 2.2*).

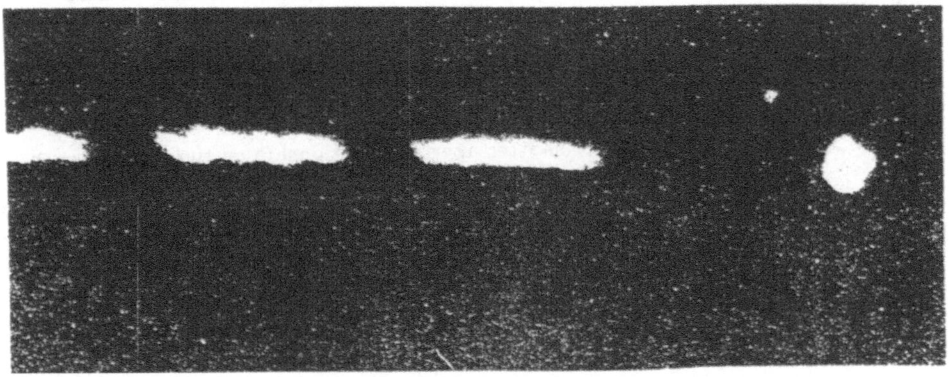

Figure 1. Positive copy. General aspect. α CVn system.

The procedure is the following: (1) We record the images of both components with telescope tracking and the right exposure time (see [*Refs. 3,4*] for an analysis of t_{exp} as a function of the magnitude). (2) We stop the tracking, switch the synchro motor on and let the image drift across the frame as long as possible. This second step records at the same time the E-W direction and the time scale.

A precision in the chopping of the trail of $\pm$ 0.05 Hz (0.1 %) with a 95 % probability that the error is below $\pm$ 0.03 Hz (0.06 %, adopted value) is obtained.

In relation to the speed of the motor, we looked for a compromise between two factors. On one side, the greater the number of segments on the trail, the smaller mean accuracy in the segment measure. On the other, few segments imply the risk that, having to skip the first segment wich corresponds to the starting of the motor and the last one (uncompleteness, deformations due to proximity to the edge of the frame) we count with only very few segments, increasing errors accordingly *(Fig. 4)* . The final speed chosen was 1 cycle every 4 seconds (*i.e.*, 0.250 cycles/sec).

The measurement of the scale factor is made by taking into account several segments (all possible but the first and last ones). This error is reduced just because the time error is known, and the error due to border estimation (bigger) is distributed over a long distance. The measurements were performed with a Nikon Metaphot microscope, with a mean precision in the micrometric movements of $\pm$ 0.005 mm at 50 X. Typical values of border estimation errors are $\pm$ 0.002 mm on a 35 mm frame. We also used a slide projector and took measures with a mi-

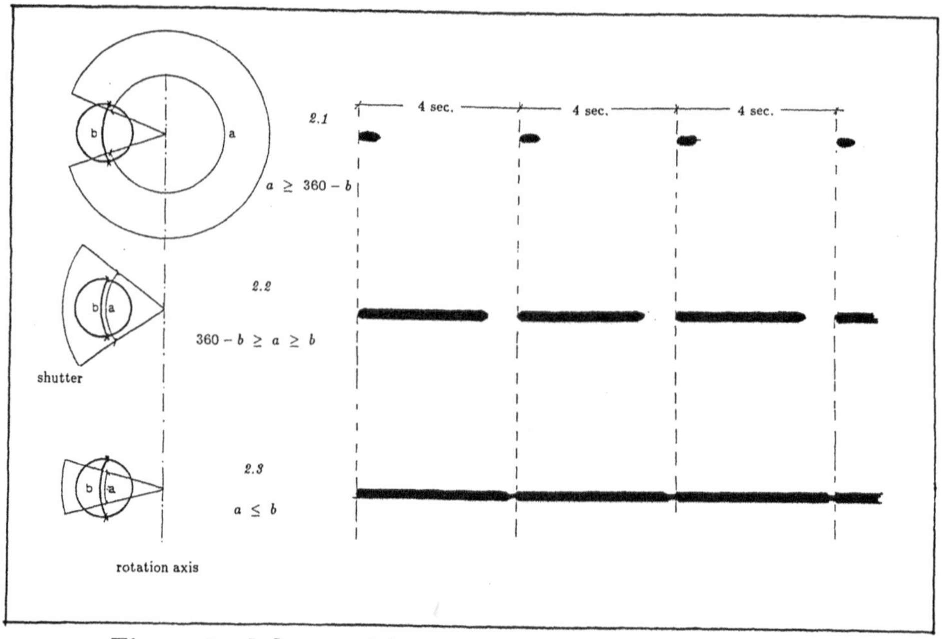

Figure 2. Influence of shutter-aperture relationship on the trace.

crometer or a good calliper. The error suffered in this process was 0.20 to 0.25 % . The final error for the scale factor measurement is typically 0.31 %.

2.3 *Measurement of separation.* At this stage, two different techniques have been employed (Table I), as illustrated in *Fig. 5*. A simple inspection indicates that the latest gives a higher error, which is consistent with the fact that it involves a quadratic operation. Typical errors at this point are 0.6 % (direct method) and 0.9 % (coordinate method). The ideal measurement consists in determining the photocentre of each component (via a microdensitometer or by image processing). The critical parameter at this stage is the exposure time (over-, under- exposition, seeing and turbulence effects, see *Ref. 5*).

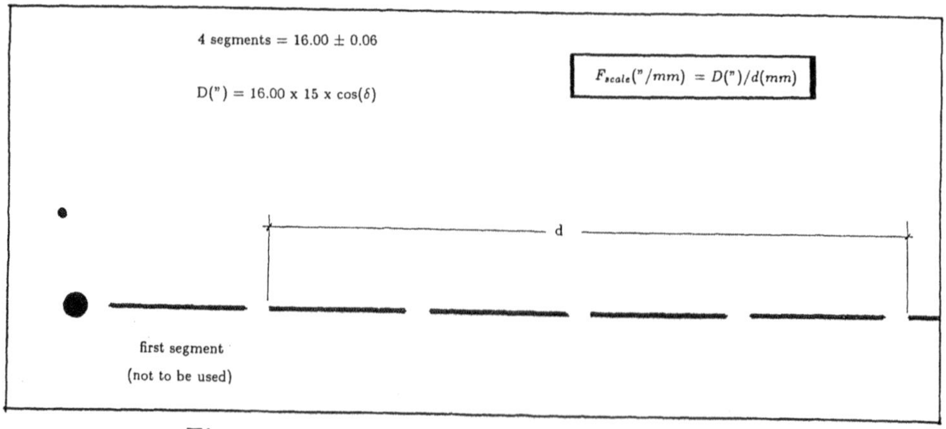

Figure 4. Measure of the scale factor. Example.

TABLE I

System	S	Esf %	$\rho_{dir}('')$	E.%	$\rho_{coord}('')$	E.%	$\theta_{arc.tan}(^\circ)$	$\theta_{arc.cos}(^\circ)$
β Cyg	6	0.31	34.61 ± 0.35	1.01	34.54 ± 0.45	1.32	53.48 ± 0.54	53.33 ± 1.21
17 Boo	1	0.49	13.65 ± 0.21	1.59	13.76 ± 0.27	2.02	235.71 ± 0.45	236.37 ± 1.72
ζ Lyr	7	0.37	44.12 ± 0.41	0.94	43.99 ± 0.52	1.12	150.14 ± 0.49	150.24 ± 0.82
α CVn	3	0.38	19.25 ± 0.22	1.17	19.06 ± 0.28	1.50	227.54 ± 0.21	226.92 ± 0.68
α Lib	4	0.24	230.41 ± 0.40	0.44	232.59 ± 1.03	0.44	313.64 ± 0.10	313.64 ± 0.18

S is the number of segments, Esf gives the error on the scale factor, and E on the preceding ρ.

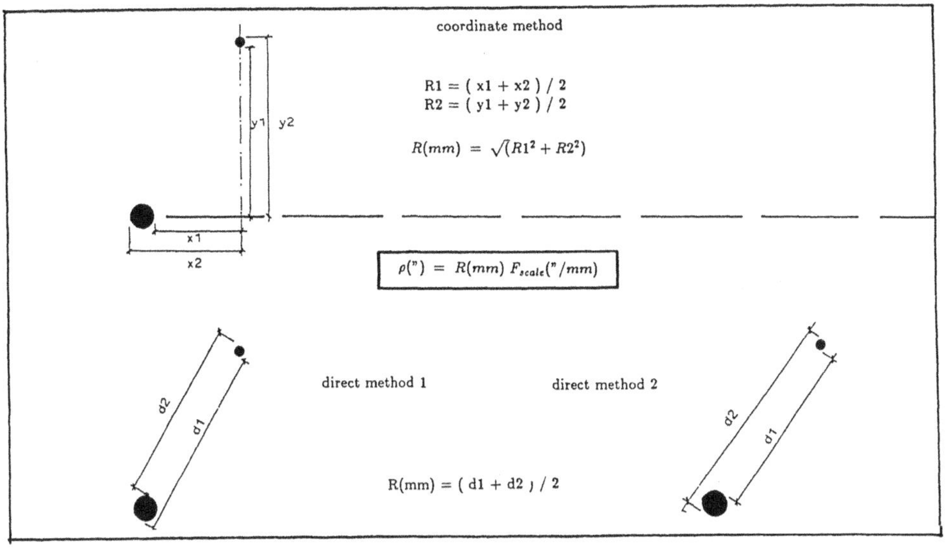

Figure 5. Measure of separation.

2.4 *Measurement of position angle.* Two different coordinate techniques may be used, one based on the use of the arc tangent and the other on the arc cosine (*Fig. 6*). For them, typical errors obtained were 18' for the first one and 50' for the second.

2.5 *Total error.* The total error has been calculated on the basis of the worst possible occurence. For the scale factor, the error is found to be around 0.24 to 0.49 %, while for the separation, 0.46 to 2.08 % (coordinate method), and 0.40 to 1.87% (direct method) are obtained. The errors for the arc tan technique are found to be 6' to 54', while the arc cosine technique gives 11' to 78' respectively. This dispersion of errors is due to their dependence on a number of factors as the border definition and background light.

3. The practice

In practice, this work has posed two main problems. The worst one was to record the chopped trail with a good definition. In fact, this problem is frequently found when using the trail method. For example, the Torino group [*Refs. 9,12*] takes series of some 20 contiguous exposures, followed by the orientation trail. In some cases, the trail is too faint and they are unable to measure the position angle. The second problem was the limiting magnitude. The image of a secondary is well recorded in exposure but is limited by the magnitude of the primary to certain values, in order to avoid large, spread images which increase the errors. The first case imposed a limit magnitude for the primary of 4.6, and in the second, a 6.7 limiting magnitude for the secondary.

TABLE II

System	ρ_{cat}	Year	θ_{cat}	ρ_{obs}	θ_{obs}
β Cyg	34.4	1967	54.1	34.6 ± 0.3	53.5 ± 0.5
17 Boo	13.5	1973	236.2	13.6 ± 0.2	235.7 ± 0.4
ς Lyr	43.7	1955	150.0	44.1 ± 0.4	150.1 ± 0.5
α CVn	19.3	1974	228.4	19.2 ± 0.2	227.5 ± 0.2
α Lib	231.	1913	314.	230.4 ± 0.9	313.6 ± 0.1

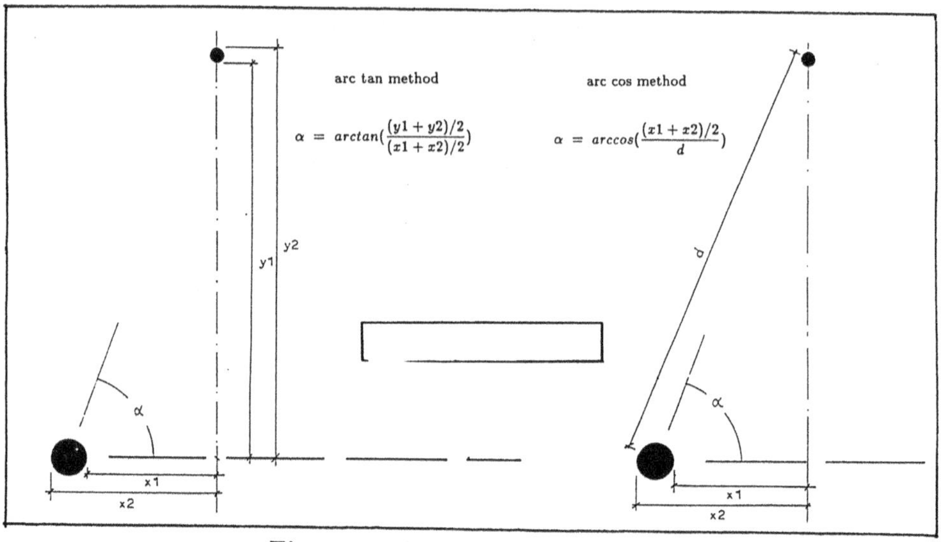

Figure 6. Measure of position angle.

We follow the simple scheme suggested by J. Dommanget [*Ref. 2*] for the data analysis. The final results are given as $<\rho> \pm \epsilon_{<\rho>}$ and $<\theta> \pm \epsilon_{<\theta>}$ where <> means a weighted average, and ϵ is the final error on the average value. The results that follow were obtained with a refractor of 90 mm, f/15, and a newtonian reflector of 200 mm, f/5, from a site near Villaviciosa de Odón, Madrid. As shown in Table II, the errors obtained are remarkably smaller than the typical visual ones [*e.g. Refs. 8,10*], and the results agree with other data [*Refs. 6,13*]. Recent photographic measures of β Cygni and α CVn [*Ref. 4*] give very similar results, even if the 0.1" error claimed in that paper seems to be quite optimistic.

4. Conclusions

The present work attempts to show that the use of a simple technique, simultaneously recording the images of components, the E-W direction and an accurate measure of time, avoids and reduces systematic errors in great proportion, besides the aditional point of keeping a record of the data.

The use of a rotating shutter has proven fundamental in the reduction of the scale factor and total errors, and its easy construction guarantees its availability to amateurs, which in this way may obtain valuable data including the accuracy range of their measures, essential for orbit determinations. The method slightly improves the accuracy of P.A., but results are not so good as for the separation. Image processing of the corresponding CCD frames reduces these errors by factors 3 to 5, depending on the signal-to-noise ratio. A rotating shutter as the proposed one remains to be developped if mounted near the focal plane of the camera. The use of this system with wider aperture telescopes will accordingly increase the limiting magnitude and image definition and will be of great assistance to the systematic measure of visual binaries.

This is an highly abridged version of our work. A complete version is available upon request.

5. References

[1] P. Couteau (1978). L'observation des étoiles doubles visuelles, pp. 67-69. Flammarion.
[2] J. Dommanget (1959). *Ann. Obs. Roy. Belgique* (Série 3) **8** : 89-93.
[3] P. Durand (1987) *in* Astronomie-Le guide de l'observateur. Vol. 2, pp.669-722. S.A.P.
[4] P. Durand, D. Simier (1987). *L'Astronomie* **101** : 265-276.
[5] W.D. Heintz (1978). Double stars. pp. 26-31. Reidel.
[6] K.G. Jones (1979). Webb Society Handbook. Volume 1. Double stars. pp. 49-58.
[7] F.J. Josties , R.S. Harrington (1984). *Astrophys. J. Suppl. Ser.* **54** : 103-113.
[8] D.J. MacConnell (1978). *Astron. J.* **83** : 436-437, and (1984) *Astron. J.* **89** : 876-878.
[9] R. Morbidelli, R. Pannunzio (1987). *Astron. Astrophys. Suppl. Ser.* **68** : 481-492.
[10] J.A. Soldevilla (1985). *Astrum* **60** : 15-16.
[11] E. Soulié (1986). *Astron. Astrophys.* **164** : 408-414 .
[12] R. Pannunzio, R. Morbidelli (1984). *Astron. Astrophys. Suppl. Ser.* **55** : 451-454.

The Scientific Merit of Amateur Astrophotography

Werner E. Celnik[1] and *Peter Riepe*[2]
[1]Astronomisches Institut, Ruhr-Universität, Postfach 102148,
D-4630 Bochum 1, Fed. Rep. of Germany
[2]Astronomische Arbeitsgemeinschaft Bochum, Alte Üminger Str. 24,
D-4630 Bochum 7, Fed. Rep. of Germany

Astrophotography is only a small fraction of the broad spectrum of amateur astronomers' work. With the development of high-sensitivity emulsions, amateur telescopes of large focal ratios and good quality, cheap multi-purpose cameras (both 35- and 70-mm), and intensive public relations work done by observatories as well as by individual, advanced amateurs, the methods and techniques of astrophotography have become accessible to all amateurs. With suitable equipment, experience, and familiarity with astronomical problems (e.g. by close contact with professional astronomers), they are even able to make valuable contributions to science.

But, unlike other fields of work (e.g. the determination of light-curves of variable stars), the scientific content of photographic images obtained by amateurs can only be extracted by using rather tricky and complicated methods, which are rarely accessible to amateurs. An important additional condition is therefore that professional astronomers be open-minded about serious work by amateurs, and also willing to support it.

Photography with cameras having large fields of view ($> 5°$) can be considered as supplementing work with professional Schmidt telescopes. Both types of instrument show similar sensitivity to low surface-brightness, found in very extended objects. Up to 25–25 mag/(arcsec)^2 can be detected by normal wide-field cameras if suitable equipment (emulsions, and photographic filters) is used. This applies to extended objects like the zodiacal light, bright comets, gaseous nebulae and dark clouds. For point sources like asteroids or variable stars, on the other hand, amateurs only have the advantage of greater observation time over professionals.

We will give two examples of cooperation between amateurs and professionals. The first one is the faint and very extended emission nebula S 27 ($> 13°$ around the central star ζ Oph (spectral type O9.5V). Figure 1 shows a wide-field image of the nebula obtained by an amateur using a normal 35-m camera, 103e-E film and a glass RG 645 filter. The exposure time was 105 minutes. The original negative film was digitized and the background, which varies over the image because of vignetting and the increasing stellar density towards the galactic equator, was subtracted. Finally the wide-field image of the nebula could be intensity calibrated in absolute physical units (cf. Celnik, Weiland, 1987). Such results can also be used for comparison with observations in other wavelengths (UV, radio), for example to determine the physical conditions within the nebula or to find out its true three-dimensional structure.

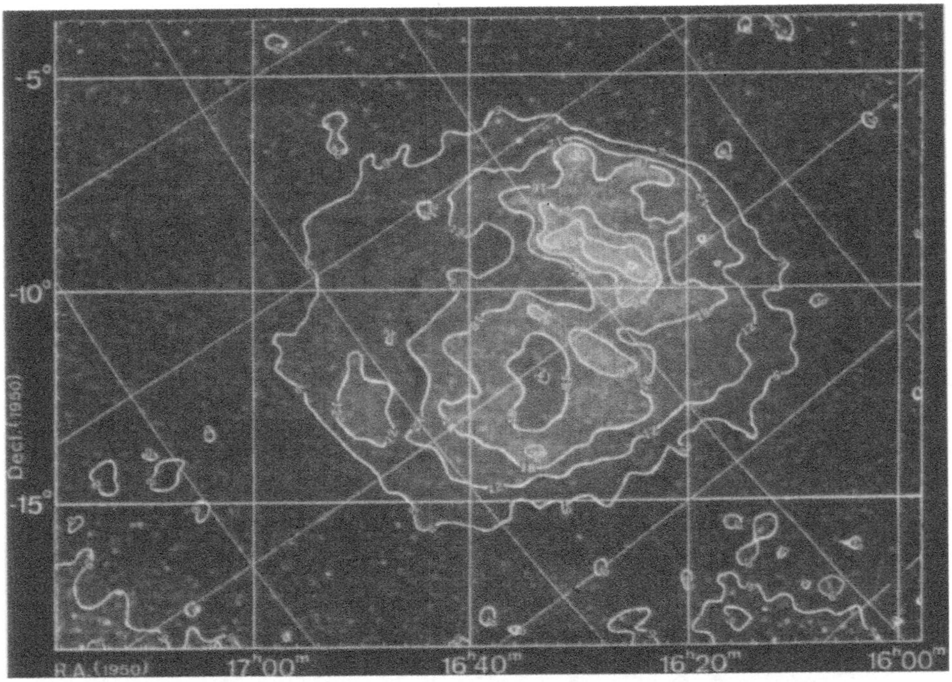

Fig. 1. H-α emission-line intensity map of the H II region S 27. The numbered isophotes indicate the density in units of 6.67×10^{-7} erg cm^{-2} s^{-1}sterad^{-1}.

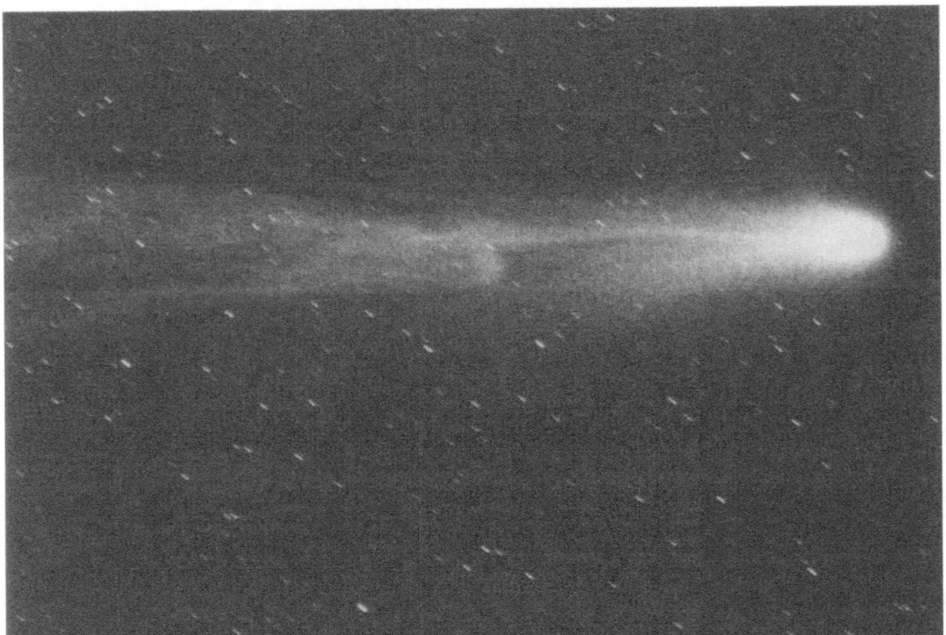

Fig. 2. Plasma tail of Comet P/Halley on 1986 March 9, 07:59 UT. A camera with a 760-mm, f/4 lens was used with Kodak 103a-F (hypered), and a filter centred on 426 nm (CO+)

The second example is Comet P/Halley, which was observed with wide-field cameras (~30° in different spectral wavelengths. Figure 2 shows the structured plasma tail of the comet. Plasma clouds with striking structure are streaming away from the coma and can be followed by photographing them hour by hour or day by day. Using quite simple mathematics, velocities (up to 300 km/s), acceleration (about 32 m/s^2), the time of zero distance from the nucleus (ejection time), and the initial velocity of each observable cloud could be obtained from many original images. In this connection the detection of faint plasma clouds far from the nucleus was very important. The distribution of ejection events in time leads to a sidereal rotation period of 50.5 ± 0.6 hrs and to the distribution of outburst sources over the surface (Fig. 3), showing two main sources at the long ends of the ellipsoidal nucleus (cf. Celnik, Schmidt-Kaler, 1987).

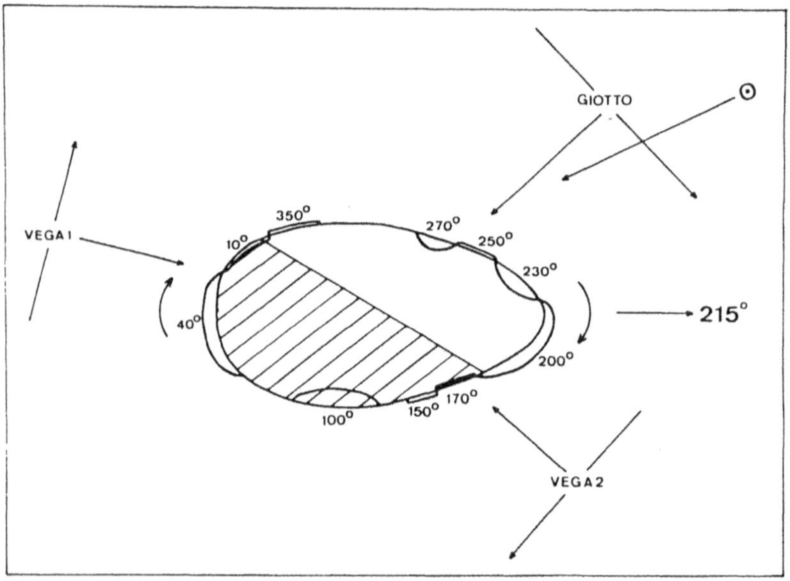

Fig. 3. Distribution of outburst sources along the equator of the rotating nucleus of P/Halley. Strong sources occur at the long ends of the nucleus. The views from different space probes during their respective fly-by encounters are indicated. The shaded area indicate the shadowed region during the GIOTTO fly-by.

The conclusion must be that amateurs are not only able to produce "nice pictures" but also that astronomical science can profit from systematic work by serious amateur astrophotographers. On the other hand this can only be successful if both amateurs and professionals are willing to cooperate.

References

Celnik, W.E., Schmidt-Kaler, Th. (1987): Astron. Astrophys. **187**, 233
Celnik, W.E., Weiland, H. (1987): Astron. Astrophys. **192**, 316

Reduction of Variable-Star Observations Using Basicode

Theo Jurriens
Kapteyn Laboratorium, NL-9700 AV Groningen, The Netherlands

The Dutch variable star section of the 'Nederlandse Vereniging voor Weer–en Sterrenkunde', founded in 1960, has about 10 to 15 regular observers. Depending on the weather conditions they perform 5 to 10 thousand observations a year. Our observation–recorder Henk Feijth collects the observations on paper and, since 1985, also on computer–cassette. From these contributions Feijth compiles a list on paper for the A.A.V.S.O. The cassette is written in a special format using the BASICODE program, see 2.

The reduction of the observations can be described as following :

1. Every observer and computer-owner enters each observation as a DATA statement in a BASIC program. So each DATA statement is a record representing one observation, each DATA statement contains 20 characters :

1- 6	Starname
7-10	int(JD-2440000)
11-13	fractional part (3 decimals)
14	: or <
15-17	estimate without decimal point
18-20	observer code

So a month of observations is simply a BASIC program consisting of just DATA statements.

2. The Dutch broadcasting organization (NOS) has designed a kind of 'Universal' BASIC called BASICODE (version 2). With the aid of BASICODE computer X can read the programs of computer Y. The observer translates his data statements (the observations) into the BASICODE format. This is saved on cassette and sent to our recorder.

3. Our recorder enters the monthly observations onto the BBC micro with the aid of the BASICODE translation program. Now the recorder runs a program that changes the star name into the Harvard number. This is done for the observations of every observer. After that the whole bunch of observations is sorted in ascending order of Harvard number and JD.

4. The file of observations obtained in 3 is used as a source for our own reports, a graphic representation of the observations, and for simulating A.A.V.S.O. record sheets. This 'end product' we send to the A.A.V.S.O.

Our system has the following advantages :

- the observer doesn't have to worry about the order of entering the observations.
- the observer doesn't need to know the Harvard number.
- the star name is entered only using capital letters the difficulty of writing Uma or UMa, etc. is avoided.
- this system saves a lot of time and allows much more possibility for reduction and improves the feedback to the observer considerably.

Detailed technical information can be obtained from the author. The set of programs for reducing the observations on the BBC micro is also available.

Image Digitizer

Pierre A. Magnien
Association Astronomique de France-Comté,
34, av. de l'Observatoire, F-25000 Besançon, France

Abstract. Equipment for digitizing 24 × 36 mm negatives was described. The sensing unit is a 1024-element CCD. Apart from the optical train, the transparency is held in a rigid carrier that is moved by a stepper motor to scan one row at a time. The whole operation is controlled by computer, and results in a 170.5 kb file for each image.

"Comet" Software

P. Butzbach and *J.P. Parisot*
Association Astronomique de France-Comté,
34, av. de l'Observatoire, F-25000 Besançon, France

Abstract. A suite of programs written in BASIC for use on Apple (II, IIe, etc.) computers was described. A data file, extracted from the general catalogue by Marsden is incorporated. The programs are applicable to any geographical location and allow the determination of positions (in both celestial and horizontal coordinates), times of rising, transit and setting, and the calculation of ephemerides from orbital elements.

Dumoulin, C.: *La Gazette d'Uranie*, Hors-Série No.1, Groupe de Recherche et d'Animation en Astronomie du Limousin, IREM de Limoges, 1986

Dumoulin, C., Parisot, J.P.: *Astronomie pratique et informatique*, Masson, Paris, 1987

Marsden, B.: *Catalogue of cometary orbits*, I.A.U., 1982

Meeus, J.: *Astronomical formulae for calculators*, Willmann-Bell, 1982

Refining the Orbit of Visual Double Stars

Edgar Soulie
"Les Dryades", 19, av. Salengro, F-92290 Chatenay-Malabry, France

Abstract. An iterative method of refining the orbital parameters of visual double stars was described. The sum of the least-square differences is minimized by the Levenberg-Marquardt method. The application to two examples was described, including one highly inclined orbit, ADS 8862 = Hussey 664 (i = 94.3 degrees).

Marquardt, D.W.: "An algorithm for least-squares estimation of non-linear parameters", *J. Soc. Industr. Appl. Math.*, **11** (2) 431–41, 1963 June

Soulié, E.J.: "L'amélioration de l'orbite d'une étoil double visuelle", *Astron. Astrophys.*, **164**, 408-14, 1986

Soulié, E.J.: Special Issue "Double Stars", *l'Astronomie*, 1988, in press

Reliability and Accuracy of Astronomical Observations by Amateurs

Zdenek Kviz
School of Physics, University of New South Wales, Sydney, NSW, Australia

Who is an amateur-astronomer? We would certainly find several definitions if we tried to get answers from the audience. I am not trying to force anyone to accept my definition, in fact I do not know if I really have one. Someone who likes to read books about astronomy? Professional scientist, expert in other areas of science, but interested in astronomy? Science fiction writer who writes about space travel? Retired professional astronomer who is no longer paid for his work in astronomy? Constructor of telescopes or astronomical instruments? Well, each of these could be discussed. But we will rely, on this occasion, on common sense and take any one who is interested in astronomy and contributes to its progress either by observation or by construction of astronomical equipment and has not an official education in astronomy. He/she simply likes the idea that he/she is contributing by his/her work to our knowledge of the universe and feels proud of it.

Obviously every amateur likes the recognition of his/her work by professional astronomers but sometimes he/she may not be aware that his/her work may be of value not only by increasing the quality of observation but also by supplying additional information about the conditions, reliability and accuracy of the observation. Any attempt to describe meticulously all conditions of observation or to check carefully the reliability and repeatability of the results may greatly improve the quality of any amateur's work. Discovery of a comet or of a nova or supernova is an important contribution itself, but can be improved with information about the estimate of brightness and limits of accuracy of this estimate. The new supernova in the LMC - a very suitable act of celebration of the anniversary of SAF - gives us a good example. New Zealand amateur Albert Jones confirming its non-visibility the night before discovery, gave an excellent limitation on the time of the supernova's outburst.

I would like to show here two examples of how accurate observations by amateurs, or at least observations made by an amateur's means can be.

Many years ago I used to organise with my colleagues visual and telescopic observations of meteors. Fig. 1. shows the luminosity function derived for both sporadic and Perseid meteors independently by three groups of eight observers. The results are plotted separately for each group and thus the scatter of the observed values gives

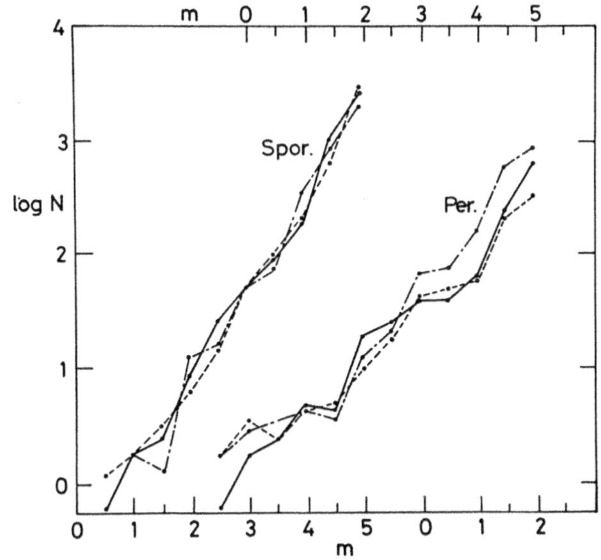

Fig.1. Luminosity function of meteors for 3 independent groups of 8 observers, separately for sporaric meteors and Perseids. Lower magnitude scale for sporadic meteors, upper scale for Perseids.

the measure of accuracy of such observations. Clearly the estimates of the brightness to one half of the magnitude are quite justified. The luminosity function was obtained using the probability of the observation of a meteor of respective brightness, the calculation of which was based on simultaneous but independent perception of a meteor by several observers. Details can be found in the original paper (Kviz 1965).

We organized various tests of observers under the supervision of a psychologist about their observational ability and reliability. Special tests concerning the alertness of the observers were also performed during the observation, to be sure that our results would be good. However, we made a mistake, we did not describe and we did not publish the precautions and tests we used. When discussing the results with a colleague I was surprised by his question "How many observers fell asleep?" He was right, how should he know anything about the reliability of our results when we did not describe the measures we did adopt? Thus I urge all amateurs to describe all details which can help someone else, who is reading the paper, to get the right idea about the reliability and accuracy of their observational results.

Now we turn to the observation of variable stars. As an example I would like to present the observation of one professional, Father O'Connell (1951), which was made by what nowadays would be called amateur methods, visual estimates of the brightness of the variable star SV Cen based on the comparison of its photographic images with the images of neighbouring stars. Fig.2. shows the comparison of the light curve based on O'Connell's normal points and the light curve from individual photoelectric observat-

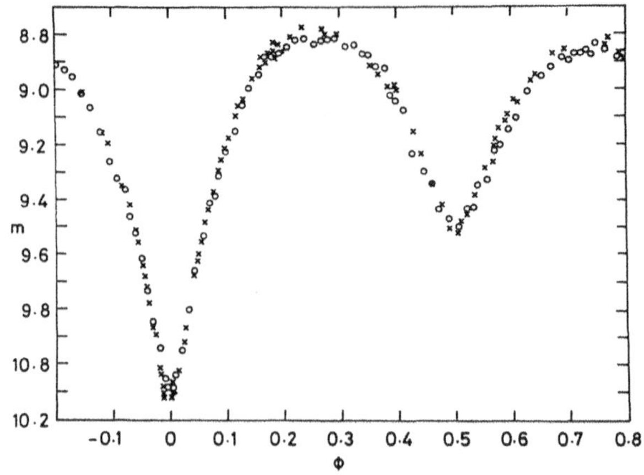

Fig. 2. Light curve of SV Centauri. o - normal points of O'Connell estimates, x - individual measurements by Irwin.

ions of the same star made by J. Irwin (1972). For those, who are not familiar with the expression "normal points", I would simply explain that each normal point represents an average of many observations (estimates). The accuracy of O'Connell's observations is widely known and perhaps very few people are capable of achieving it, however this example shows clearly what any amateur can do if he/she is experienced, patient and diligent with a small photographic camera. These days many amateurs can afford even photoelectric photometers and proportionally better results can now be achieved with less effort. However, we should keep in mind that photographic records have still a great value as they contain information about many stars and will be even more valuable for future generations of astronomers tracing the past changes of various objects in the sky.

With observations of professional astronomers it is tacitly assumed, that all precautions concerning the proper functioning of the equipment, including the clock, and all necessary corrections were taken into account. It is considered as a professional responsibility of the author to guarantee the reliability of the results. The amatuer, espacially if he/she is not yet well known for his/her work, should put all effort into describing in detail how the observations or measurements have been done. (Sometimes even professional astronomers omit important information concerning the methods used). Take, for example, timing of the minima of eclipsing variable stars. Often there are no regular observations, there are gaps in measurements, etc. Now, putting all times of minima together we may see that one minimum, perhaps quite isolated, does not fit among the others. What is the reason? Real jumps in the period changes of the star? Malfunctioning of the clock? Omission to take into account the heliocentric correction? What should an astronomer studying period changes of such a star - perhaps many years later, when the observer is no longer alive to be consulted - think about that? Please, preserve the proper description of your observations for future use.

The amateurs, as the word itself, based on its Latin origin "amare - to love" indicates, love astronomy, they are not paid for their work. Some of them, as the discoverers of comets, are honoured in the way that the comets are named after them. I wish to propose, that discoverers of novae, supernovae and other variable stars should also be honoured in similar way, the stars should bear their names. There are many stars and many problems about them need to be solved. Professional astronomers having now very powerful telescopes in all spectral ranges of electromagnetic radiation are too busy with a great variety of new exciting objects. Many survey programs of professional observatories have been discontinued. By encouraging the amateurs in this way, we could gain much more information about various astronomical objects and we would be praised for that by future generations.

References:

Kviz 1965, BAC 16, 263.

O'Connell 1951, Riverview Coll. Obs. Pub. 11, 69.

Irwin J.B. & Landolt A.U. 1972. PASP 84, 686.

Trying to Reduce Errors in Visual Estimates of Variable Star Magnitudes

E. Velasco and *P. Velasco*
Agrupación Astronómica de Madrid, Apartado 1.039, Madrid, E-28080, Spain

Introduction

The light-curves of many variables, prepared from a large number of individual estimates by many different observers, have a lot of scatter, ranges greater than one magnitude being common, although it is generally accepted that the eye can detect differences of roughly 0.1 magnitude, at least for stars of moderate brightness. The question arises whether these gross discrepancies can be reduced. We think that this is possible if observers make better use of their observational capabilities.

For certain types of variable stars, e.g. long-period ones, the lack of accuracy may often be balanced by the vast number of estimates available. But for other stars, especially those exhibiting small-amplitude fluctuations, this process of combining observations leads to broad light curves that completely hide any possible fluctuations. Variable-star observers are familiar with these problems, but it is desirable to try to improve the accuracy that can be obtained.

Observational Errors

Apart from purely accidental errors, such as wrong identification of the variable or of one or more comparison stars, there are several sources of error that affect the quality of any visual estimate. These fall into two categories: random errors, which are present in any scientific measurement, and systematic errors, introduced by the measuring device, in this case telescope plus observer.

Random errors tend to cancel out as one considers more and more independent measurements. The r.m.s. deviation is $1/\sqrt{(N)}$, N being the number of measurements made. The origin of systematic errors is less evident, and its reduction depends on a thorough study of the process of estimation. A further difficulty is introduced by the bias on the part of the observer.

These errors can be grouped as follows:

1. observers make estimates with different telescopes (different aperture, power, focal length, eyepiece),
2. observing conditions are different, from dark, clear skies to light-polluted ones,
3. observers use different comparison stars, or different charts with different magnitudes for the stars, and
4. the thoroughness, experience and sensitivity – especially to colour – of observers are not the same.

The most important errors come from points 3 and 4, and they can be removed or at least reduced by using additional pairs of comparison stars, re-assessment of the steps assigned to comparisons, and improvement of charts, along with the usual observational precautions.

Study of Errors

Additional Pairs. The Argelander method is the most popular for estimating magnitudes. By assigning a step to each comparison star in a pair that brackets the brightness of the variable, the magnitude of the latter can be inferred. The step is chosen from a universal scale of steps.

An obvious extension of the Argelander method, sometimes mentioned in the literature but not sufficiently stressed, is the use of additional pairs of comparisons. More observers now use this method because it leads to an improvement in the accuracy by reducing random errors, with little increase in effort and observing time. A more accurate value is obtained by averaging results for each pair of comparisons. This is apparent in the light-curve of SS Cygni in Fig. 1, where dots represent averaged results, but individual estimates lie within the error bars.

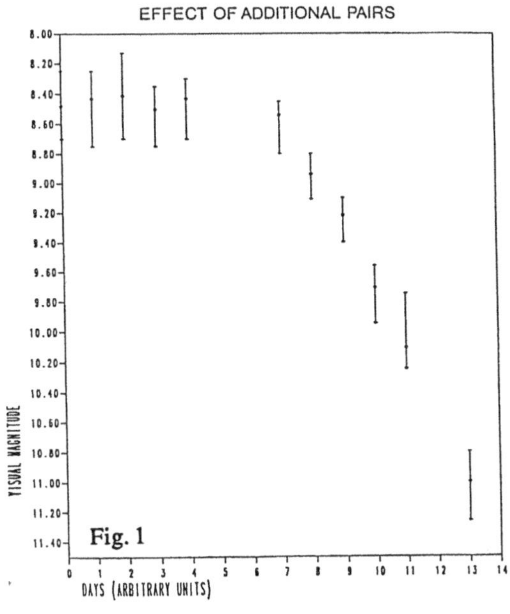

Fig. 1

In the light curve of Fig. 2, important oscillations seen by one observer, who used additional pairs, are missing for another person who did not. Another good light-curve, that of the 1985 fade of R Coronae Borealis, is shown in Fig. 3. Two observers were involved in the measurements, and small-scale departures from the mean curve could be detected.

But not only are random errors reduced with this practice. Sensitivity problems, for example if any of the comparison stars is red, are also diminished, as well as problems introduced by erroneous magnitudes.

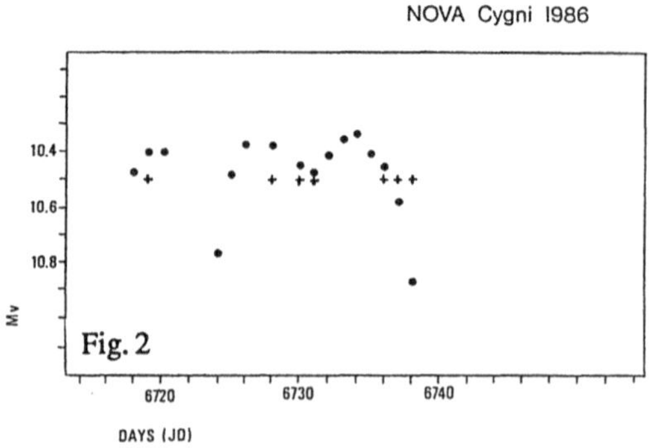

Fig. 2

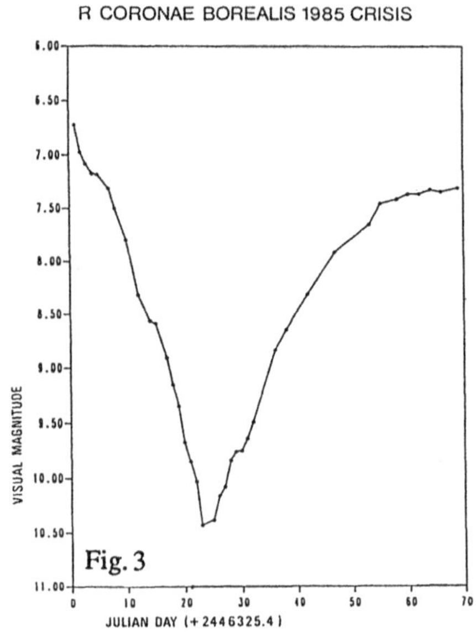

Fig. 3

Re-Assessment of Steps. The Argelander scale of steps is, in principle, universal, but each observer applies it in a different way. No problem arises if this personal interpretation amounts to a simple renormalisation of all steps. However, an observer may like to give large steps only when the difference in brightness between a comparison star and the variable is very large, but that may not apply to other observers. The result is that for that particular observer the estimated magnitude will be lower or greater than the average.

We have devised a simple procedure for performing this personal re-assessment of steps, and we think that every observer should do likewise, looking for consistency between his scale of steps and the scale of magnitudes, which are assumed to be linearly related in the Argelander interpolation formula. If this consistency is not fulfilled, bad results will be obtained.

Comparison Star Charts. The use of comparison star charts with wrong magnitudes contributes in an important way to obtaining false light-curves. Observers often find inconsistencies between magnitudes and observations. This is not strange taking into account that sometimes they use charts drawn decades ago. When inconsistencies appear, and lacking recent photoelectric measurements, estimates of the magnitudes of the comparison stars should be made visually. This can be done from any two or more comparison stars whose magnitudes are known. When measurements from different observing sessions are averaged and the new magnitudes used from then on, better individual light-curves are obtained simply because compatibility between the observer's eye and magnitudes has been reached. For an observer sensitive to red, the use of a red comparison star makes the variable appear fainter than it really is, and therefore it seems convenient to use a magnitude for that comparison star that agrees with what the observer sees. From this point of view each observer is a particular "photometer" and the question arises as to whether it makes sense, strictly speaking, to collect data from slightly different "photometers"

Limiting Magnitude. Finally, there are the effects related to different instruments and observing conditions. These effects are less important than those just mentioned. Some time ago, however, we and other members of our astronomical association observed a possible aperture-dependence. Such a dependence seems established for comets, thanks to the work of Bobrovnikoff and Morris. (Earlier in this Colloquium, however, S. Edberg pointed out some possible flaws in this aperture-correction for comets). Recent studies by Bateson, presented at the 46th Colloquium of the IAU in 1979, did not confirm this for variables.

We undertook a study with telescopes of different apertures. The observing method included the techniques outlined above and also the out-of-focus technique, which is useful in estimating stars that are close in brightness or red. To avoid position-angle problems as much as possible, it is also good observational practice to observe from one side of the telescope and then the other – if using a reflector – to compensate for the uneven response of the retina. The results of our work are shown in Fig. 4. Borrowing an idea from Bobrovnikoff, a linear relation between observed magnitude and aperture was assumed,

$$m_{\rm obs} = m_{\rm stand} + \alpha(A - 5)$$

where $m_{\rm obs}$ is the estimated magnitude, $m_{\rm stand}$ is the standard magnitude, defined to be the magnitude corresponding to a previously chosen standard aperture, which in this case was 5 cm. A is the aperture of the instrument in cm. The parameter α is obtained from a least-squares fit. The Figure plots magnitude correction – that is, estimated magnitude minus standard magnitude – versus aperture. A definite trend is apparent: the larger the telescope, the fainter the star. This diagram incorporated several hundred observations made by the authors, spanning about a year. Some observations were taken from the authors' observing programme, not at first intended for use in this particular study. We ruled out red stars in order to avoid colour problems, and used the observational techniques quoted above.

However, we think that a more sensible parameter is the limiting magnitude of the telescope, which depends on both aperture and observing conditions, thus

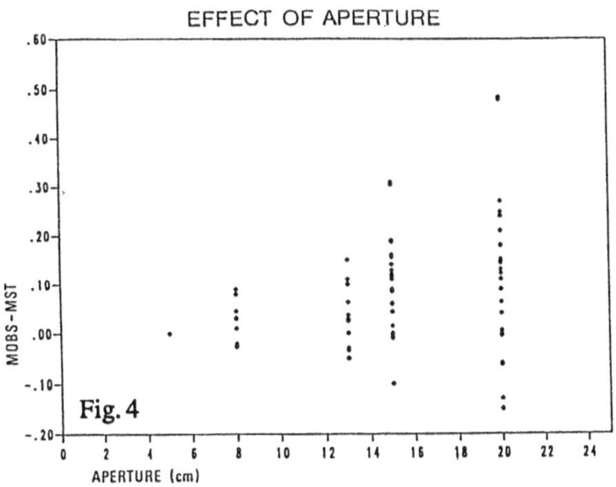

Fig. 4

including several factors in a single parameter. The limiting magnitude is a barrier near which stars are more difficult to observe, but at the same time differences in brightness are more easily detected. Much brighter stars may not be so clearly distinguished. This effect can be seen in Fig. 5, which plots Argelander step versus distance [$= \Delta m$ – Eds.] from limiting magnitude. For a given difference in brightness between a pair of stars, say 0.5 magnitude, the step assigned to this separation tends to be higher as we approach the limiting magnitude. In theory, at the limiting magnitude, the step would be infinite if one of the stars were to lie on the border and the other below it. Furthermore, a straight line with non-zero slope is seen for values less than four magnitudes above the limiting magnitude, while a plateau is reached for values greater than four.

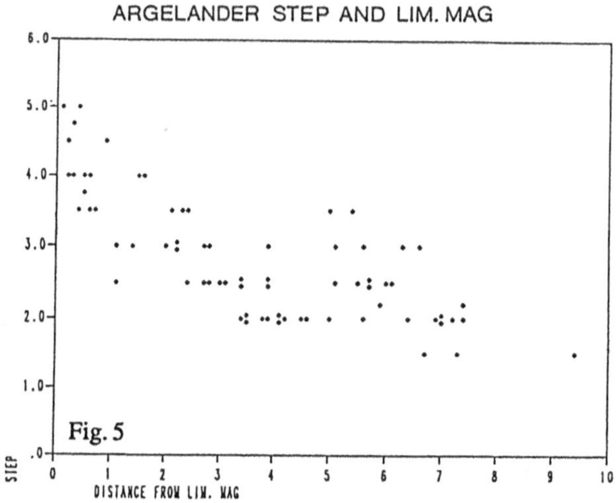

Fig. 5

We are continuing research along these lines in order to see if quantitative information, useful in the reduction of data, can be extracted.

Conclusions

In summary, we would advise observers to make an effort to imporve their estimates. Only in this way will large-scale collection of data be more significant. We speculated that a possible correlation existed between instrumentation plus observing conditions, and observational errors. Careful visual measurements were made to check this hypothesis, but no distinction was made between type of telescope, i.e. reflector or refractor, because we wanted to emphasise the important role played by the limiting magnitude parameter. Special care was taken with the colour of the stars, an aspect that was outside the scope of our study. This work is not free from shortcomings, of course, such as poor statistics, and further study will be needed to clarify the issue.

The authors acknowledge the members of the Agrupación Astronómica de Madrid for general support, especially Jos Ripero for useful comments, Jos Prieto for instrumental help, and Jesús Angel Rodrǵuez and Diego Rodrǵuez for assistance with the literature on the subject.

Part III

Observations and Results

Photographic Observation of Stellar Occultations

A. Tomić[1], *N. Čabrić*[1] and *V. Čelebonović*[2]
[1] Narodna Opservatorija, Kalemegdan, 11000 Beograd, Yugoslavia
[2] Institute of Physics, POB 57,11001 Beograd, Yugoslavia

Stellar occultations are usually observed visually, but this can also be done photographically. Our method has some advantages :

- the obtained moments of contacts refer to the smoothed lunar limb,
- the error in the determination of contacts is calculable,
- the error may be less than in visual work,
- the reduction of photographs is fast, due to the application of an "on–line" computer [1].

In order to achieve all this, it is necessary to make a series of photographs of the Moon and the object being occulted before the first and after the last contact. This is a crucial step in our method, because the photographs must be of high quality. This demands two things :

- the images of the occulted object and the Moon must be reliably recorded and
- the illuminated limb of the Moon must give optimal darkening on the film.

The required exposure time is given by [2] :

$$t_M = \frac{(F/D)^2}{\tau_o S} \frac{1}{1600\tau_a \, A \, b(\lambda, \psi)} \tag{1}$$

and for the star :

$$t_S = \frac{(F/D)^2}{\tau_o S} (C_o C_S / LF)^2 \, 10^{(0.4m+0.67)} \tag{2}$$

where both time intervals are in seconds, D is the diameter of the objective whose focal length is F, τ_o and τ_a are the transmitances of the optical equipment and the atmosphere (≤ 1), S is the sensitivity of the photographic emulsion (ASA), A is the albedo, m the photographic stellar magnitude, $b(\lambda, \psi)$ is the distribution function of the brigthness on the Moon (λ the selenographic longitude and ψ the lunar phase angle), L(N) is the resolving power of the objective (film) (mm^{-1}), $C_S = 1 + L/N$ characterizes the reduction of the resolving power, and C_o is a numerical parameter, describing the atmospheric spreading (Δ") of the point image : $C_o = \Delta/(1+kD/FN)$. For a panchromatic film, $k = 1500 \ mm^{-1}$.

Therefore, the exposure time is adjusted to the Moon, and the continuously variable parameter will be the age of the Moon. It is important to determine the limiting stellar magnitude attainable. It follows from eqs. (1) and (2) that :

$$m = 1.875 - 2.5\, log(A\ b(\lambda,\psi)) - 2.5\, log(\tau_a C_o^2\ \delta^2(1 + kD/FN)^2) \tag{3}$$

where δ is radius of the Airy disk.

The limiting magnitude increases with increase in D/F, S, τ_a, C_o. At the same time the exposure time becomes shorter. A greater value of N gives a larger m, but it demands longer exposure. Worse observing conditions (larger C_o) increase the exposure time and reduce m.

An example, calculated for a Zeiss 110/2000-mm refractor, with S = 32, τ_o = 0.8, τ_a = 0.5, C_o = 1 is shown in the figure. Note that the point d = 14 corresponds to New Moon.

Table I – Some values of the function b(90, ϕ)

age (days)	1	4	8	12	16	20	24	28
b(90,ψ)	0.02	0.34	0.69	0.94	0.97	0.81	0.49	0.08

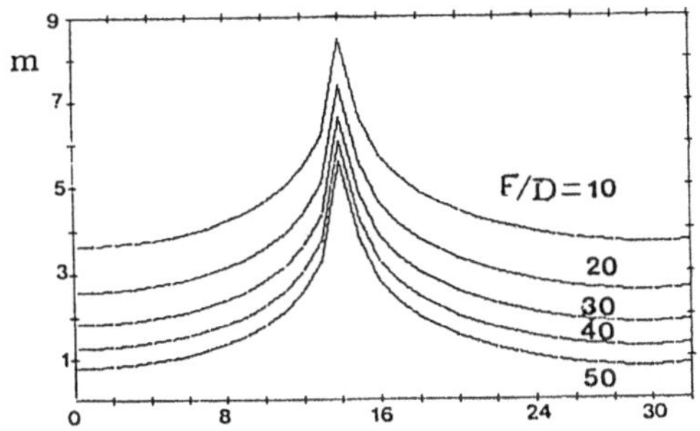

References :

1. Čabrić, N., Tomić, A. and Čelebonović, V., 1985, *Publ. Obs. Astron. Belgrade*, 33, 75.
2. Tomić, A., 1983, *Astrophotography*, Astron. Soc. of the University, Sarajevo, Yugoslavia (in Serbo-croat).

Studying "Luna Incognita": The Region Near the South Pole

Michel Legrand
1 rue de la Fontaine, F-35134 Coesmes, France

Abstract

A region of approximately 270 000 km^2 near the south pole of the Moon has not been mapped by spacecraft and Dr John Westfall of ALPO proposed the "Luna Incognita" programme in 1972 to try to cover this area. A brief summary of the problems of observing this limb region was given, together with the author's experience using the T 60 and 1-m telescopes at Pic du Midi.

References

J. Assoc. Lunar Plan. Observers: **29** (9–10), pp.180–3, 1982 Dec.; **30** (3–4), pp.61–5, 1983 Oct.; **30** (9–10), pp.199–205, 1984 Aug.; **30** (11–12), pp.221–6, 1984 Nov.; **31** (5–6), pp.119–23, 1986 Jan.; **31** (11–12), pp.258–60, 1986 Nov.

Sky & Telescope, 1984 Mar. pp.284-6

Westfall, J.E.: "Lunar mapping", in *Observing the Moon, Planets and Comets*, Chapman, C.R. & Cruikshank, D.P., eds., Vol.2, pp.181-207

Audouze, J., Israel, G., eds.: Cambridge Atlas of Astronomy, Cambridge & Feltham, Mddx., 1985

Neel, R.: "Pour une carte de France de la turbulence", "Ciel et Espace", (205) p.49, 1985 May/June

Rükl, A.: Moon, Mars and Venus, London, 1976

Asteroid and Cometary Occultations

David W. Dunham
P.O. Box 7488, Silver Spring, MD 20907, U.S.A.

Occultations of stars by minor planets and comets can be used to determine the diameter of the occulting body. With photoelectric equipment, it might also be possible to measure the diameter of the occulted star or to resolve close binary systems at the same time.

On 1983 May 29, (2) Pallas occulted the 4.7-mag. star 1 Vulpeculae. A star this bright is occulted by one of the four largest asteroids only once every 60 years. Timings were sent by 130 observers in Florida, Louisiana, Texas, Arizona, Sonora, and Baja California, making it the best-observed asteroidal occultation to date. Asteroidal grazing phenomena were confirmed for the first time. The star is a spectroscopic binary. Several observers timed contacts of the 6th-mag. companion, determining the separation to ± 0.0004 arcsec. An elliptical fit to the timings projected onto the plane of the sky has been made. Unfortunately, clouds prevented timings from near the southern limit, so about 100° of Pallas' circumference was not observed.

The occultation of 14 Piscium by (51) Nemausa on 1983 September 11 was observed from approximately 50 stations from south-eastern Alabama to south-eastern Virginia. Photoelectric observations were made with portable 36-cm Schmidt-Cassegrain telescopes operated by Ted Dunham and Richard Baron (MIT) at sites near Emporia, VA, and in eastern NC. Another portable photoelectric station recorded the occultation at the Mark Smith Planetarium at Macon, GA; Glenn Schneider of the University of Florida directed the effort. A fourth photoelectric record was obtained at the NASA Langley Research Center, Hampton, VA. Joan and David Dunham obtained a television record at Essex Meadows, VA a southern suburb of Norfolk.

Video-records of close approaches to 6th-mag. stars by Comets Giacobini-Zinner and Halley showed no dimmings by cometary matter. Theoretical models show that material dense enough to cause noticeable dimming is usually present only within 10 km of the nucleus. Astrometric errors were large, so nearly 100 stations were needed for confirmed timings from two or more sites. During a close Halley appulse in November, observers at separate sites near Brisbane timed similar large dimmings for half a minute. Large amounts of dust may be released sporadically, causing dimmings over large distances. The 5-fold difference in dust detected by the two VEGA spacecraft seems to confirm this.

Another occultation of a 8.7 mag. star, SAO 104751 by Pallas on 1983 May 4 should be mentioned. A 23.49-second occultation was recorded photoelectrically at the Engelhardt Observatory in Kazan (USSR), the only observation of this event. Unlike the condition for any of the May 29th observers, the sky transparency was good and conditions were photometric at Kazan on May 4th. A report was published by V.B Kapkon in *Soviet Astronomy Letters*, **10**, p.26 of the English translation (p.67 of the original, edition of 1984 Jan.–Feb.). The observations show a dimming before and after the occultation. Kapkov has prepared a computer-generated model, and suggests that the dimming may be due to a cloud of dust grains and gas resulting from meteorite impacts on Pallas.

References

Dunham, D.: "Preliminary Report on the Pallas Occultation", Occultation Newsletter, **3**, (2) pp.76–7 (1983 July)

Dunham, D.: "Observations of Appulses of Halley's Comet", Occultation Newsletter, **3**, (16) pp. 352–4 (1986 June)

Dunham, D., Dunham, J., Anderson, P.: "Asteroidal and Cometary Occultations", *Proc. Astr. League 40th Ann. Mtg*, Baltimore, 1986

Note: The results of the Pallas 1983 May occultation of 1 Vulpeculae have been submitted for publication in *Astron. J.*

Astrometric Measurements of Rarely-Observed Minor Planets

Roger Chanal
Nurol, F-43110 Aurec sur Loire, France

Abstract

The 1986 Minor Planet Ephemerides call observers' attention to the need for astrometric observations of 120 objects on the "critical list". Their magnitudes are generally between 15 and 18 at opposition. The author's programme of observation with simple means was described. The equipment used was a Newtonian 41-cm, f/4.8 reflector, a standard camera back, and Kodak TP 2415 film, hypersensitized in forming gas. Out of 25 observations attempted, 20 photographs were obtained, 8 of which were of sufficient quality for the positions to be measured and to be published in *Minor Planet Circulars*. On two photographs, the minor planet (1982 KM) was detected, but there were no suitable reference stars within the field. (The lack of reference stars brighter than mag. 10 in certain regions should be noted.) The return of measured positions versus photographs obtained (40%) appears very acceptable.

Video Recordings of Asteroidal Occultations

Roland Boninsegna[1] and *Jean Lecacheux*[2]
[1] rue de Mariembourg, 33, B-6381 Dourbes, Belgium
[2] Observatoire de Paris, F-92195 Meudon Principal Cedex, France

Abstract

Text not received

Video recordings of two occultations were shown, the first concerning the star AGK3+29 0398 and (93) Minerva on 1982 November 22, 03:31 UT, using the 1-m telescope at Pic du Midi. The occultation was also observed in Spain and in the U.S.A. The second event involved AGK3+17 1309 and (146) Lucina on 1982 April 18, 20:23 UT, using the 1-m telescope at Meudon. A short secondary occultation (0.6 sec) was observed, not caused by the minor planet. A possible interpretation is a small satellite (5.7 km minimum diameter) near Lucina.

References

Arlot, J.-E., et al.: Icarus, **61**, 224, 1985
Millis, R.-L., et al.: Icarus, **61**, 124, 1985
Arlot, J.-E., et al.: Icarus, **61**, 224, 1985

Space Exploration of Comets and the Evolution of Amateur-Professional Cooperation

A.-C. Levasseur-Regourd
Université Paris IV, Service d'Aéronomie, BP.3,
F-91371 Verrières le Buisson, France

Introduction

In 1985–86 when spaceprobes flew past P/Giacobini-Zinner and P/Halley, many amateurs observed these comets and submitted their results to national or international (IHW) organisations. Their visual, photographic, photometric and spectroscopic observations made a contribution to scientific knowledge. It is pertinent to ask if space programmes under consideration will allow similar cooperation to be continued.

Possibilities for the Exploration of Small Bodies in the Solar System

The least uncertain project is that of another cometary fly-past by GIOTTO. Following orbital corrections made between 1986 March 19 and April 2, ($\Delta V = 115\,\mathrm{ms}^{-1}$), the European probe should pass 22 300 km from the Earth on 1990 July 2. The gravitational assist from our planet should then put it onto an orbit to pass P/Hartley 2, P/du Toit-Hartley or P/Grigg-Skjellerup (1992 July 10). This last comet appears to be the most desirable target; although the final decision will not be taken until autumn 1989, recommendations have been made (Morley, 1986) for an international astrometric campaign.

There is also a lot of discussion about the American CRAF (Cometary Rendez-vous Asteroid Fly-by) mission. NASA envisages using a spacecraft of the new Mariner Mark II series to follow the changes in activity of P/Wild 2, P/Kopff or, more probably, P/Tempel 2. Launch will take place in 1993 February; after using the gravitational assistance of the Earth and Mars, and passing the minor planet Hestia, the rendez-vous will last from 1996 to 1999 (Neugebauer, 1987). CRAF has not yet received official approval; but its experiments have been chosen, and a call for astrometric, photometric and spectroscopic observations have been made.

Other missions are still being studied. The American GALILEO probe, grounded by the Space Shuttle disaster, may fly past a minor planet before arriving at Jupiter. The VESTA Franco-Soviet-European vehicles (Perret, 1987) could fly past 5 minor planets and one comet (P/du Toit-Neujmin-Delporte?), between 1994 and 2000. Finally, there is the possibility, at the beginning of the 21st century, of a CNSR (Comet Nucleus Sample Return) mission, with the return of samples obtained either from the coma or better, from the nucleus of a comet (Schwehm, 1986; Eberhardt, 1986).

Comets Grigg-Skjellerup and Tempel 2

It is therefore these two comets that are of interest to space agencies. They are short-period objects belonging to Jupiter's family. They are relatively old, low in dust, faint, and thus quite different to Halley. P/Grigg-Skjellerup passed perihelion on 1987 June 18 (magnitude $\approx$ 13) – specially for this Colloquium – and P/Tempel 2 in 1988 September (magnitude $\approx$ 10) and again in 1993 December (Table 1).

Table 1. Orbital elements (Marsden, 1986)

Comet	T	q(AU)	e	i(°)	Ω(°)	ω(°)	P(yrs)	NGF
G.-S.	1982 May 15	0.989	0.666	21.1	212.6	359.3	5.09	+0.01; -0.0011
Tempel 2	1983 June 1	1.381	0.545	12.4	119.2	190.9	5.29	+0.05; +0.0016

The success of future missions depends on better knowledge of the evolution of orbits (the problem of non-gravitational forces) and of both the orientation and the rotational period of the nucleus (position of active areas); which explains the importance of observations made by professionals and advanced amateurs.

The Role of Amateur Astronomers

Amateurs that have been trained in rigorous methods and having access to powerful instruments have an essential role in astrometry and photometry of comets (or minor planets) that may be explored between now and the end of the century. The computing facilities that most amateurs possess allow them to carry out preliminary analysis of their observations (correlation with ephemerides, search for periodicity, etc.). But the work of amateurs who just observe minor planets, meteor showers, or "good" comets, although seemingly less important, should not be under-estimated. The space exploration of the Solar System represents a "new frontier", in which amateurs, with their love and curiosity for the sky, should be the first to participate.

References

Eberhardt, P. et al.: in *Exploration of Halley's Comet*, ESA SP 250, 243, 1986
Marsden, B.A.: *Catalogue of Cometary Orbits*, IAU, 1986
Morley, T.A.: *GIOTTO study note 61*, ESOC, 1986
Neugebauer, M.: in *Diversity and similarity of comets*, ESA SP 278, in press, 1987
Perret, A.: *Rapport de phase A Vesta*, CNES, 1987
Schwehm, G.H., ed.: *Comet nucleus sample return*, ESA SP 249, 1986

The Contribution of Amateur Astronomers to Cometary Observation

André Heck
Centre de Données Stellaires, Observatoire Astronomique, 11 rue de l'Université, F-67000 Strasbourg, France

Most professional astronomers do not conduct specific comet-discovery programmes and a new comet is, in general, an intrusion into an observational programme, although one that does, of course, receive all the attention it deserves. Comet discoveries are by-products of professional work. This may also be the case with some amateurs (G.E. Alcock discovered his fifth comet while looking for novae). Most amateur discoveries are the result of active and systematic work. Over the period 1951–70, on average 9 comets per year were discovered by amateurs.

The time spent scanning the sky before discovering a comet is typically hundreds of hours. The times of W. Bradfield between his first comets were 260, 306, 145 and 106 hours. Astronomers at Skalnaté Pleso discovered eight comets in six years, averaging one comet for 206 hours of searching. R.D. Austin and R. Meier spent 151 and 105 hours, respectively, between discoveries. These figures are only indicators, because many other factors (including luck) play a part. Alcock's two comets in six days is quite exceptional.

It is essential that any possible discovery should be carefully checked and repeated at least an hour, or preferably a day, later, so that any motion may be apparent. Photographic observers should interrupt their exposures asymmetrically in order to show the direction of motion. A positive finding should be confirmed by another observer before announcement. The IAU *Circulars* should be checked, if necessary by contacting a professional institute (for useful addresses see Heck & Manfroid, 1986, 1987). The Central Bureau for Astronomical Telegrams may then be informed.

In recent decades there has been less and less professional astrometry of comets. The Central Bureau therefore welcomes accurate measurements and reductions from amateurs.

At various times there have been prizes for comet discoveries, but there is none at present for European amateurs as there is, for example, in the U.S.A. Can no society or company be encouraged to set up such a prize for Old-World astronomers? It would be an excellent incentive to European amateurs.

References

Heck, A.: *L'Astronomie*, **88**, 318 (1974)
Heck, A.: *Ciel et Terre*, **101**, 115 (1975)
Heck, A., Manfroid, J.: *International Directory of Professional Astronomical Institutions – IDPAI 1987*, (1st edn), CDS Spec. Publ. **9** (1986)
Heck, A., Manfroid, J.: *International Directory of Astronomical Associations and Societies – IDAAS 1988*, (7th edn), CDS Spec. Publ. **10** (1987)

New Method of Statistical Analysis of Cometary Light-Curves Applied to P/Halley

Serge Thebault
122 rue de Général-Galliéni, F-92100 Boulogne, France

Various methods of determining the rotation of P/Halley's nucleus give 2 periods: 53 hours, and 7.4 days (1). Cometary light-curves are commonly a function of heliocentric magnitude versus the logarithm of heliocentric distance: the result is a set of scattered points. The average light-curve is calculated by the least-squares method. The method proposed here is a reduction of scatter by computation to derive the comet's specific light variations. The factors causing scatter are: observer experience (physiological and psychological effects), instrumental factors, comet morphology, methods of comparison, and observing conditions. Recently Morris and Edberg (2) have shown that the first three of these were the most important factors. In fact, the prime necessity is to consider visual magnitude estimates as a statistical sample with variability caused by scattering factors. This effect is different for each sample of estimates. Mathematical correction will be applied only after a statistical analysis of the effect of the factors causing error. The method consists of four steps:

1) Computation of heliocentric magnitude (mH) and study of the number of observations for each observer. Observers with less than 10 are omitted (because the least-squares method is not reliable in such a case).

2) Correction of instrumental factors and observer effect. Computation by least-squares method is carried out between (mH) versus instrument diameter (D) and exit pupil (P). The diameter and exit-pupil correction is made only if the correlation is significant and not systematic. The correction for observer effect is achieved (for each observer) by least-square computation between mH and logarithm of heliocentric distance. The O−C magnitude is plotted on a graph and a preliminary curve drawn.

3) Correction of bias: when a magnitude estimate is made on one night, an estimate the next night will tend to be equal because of psychological effects. It is possible to detect this effect by comparing the number of equal and different magnitude estimates (±0.1 mag) made at time-intervals of 24 hours, and over 24 hours. This comparison is made with the χ^2 test. If the test is significant the biased estimates are omitted.

4) Other factors (systematic error, photoelectric sequence of comparison stars) are detected by multi-factorial analysis and deleted.

The light variations of P/Halley have been studied by using 504 visual estimates reported in *International Comet Quarterly*, for 1985 Aug.–Oct., only 390 providing a database for computing the preliminary light-curve. The distribution of instrumental factors shows that diameter and exit-pupil effects are not significant during August and September ($P > 0.05$). During October the instrumental effect is significant ($P < 0.05$) and corrections have been applied to this month only. The study of bias has shown that estimates made 24 hours apart are equal in 91% of the data and different in 9%. Estimates made at intervals of 48 hours and more are equal in 59% of data. The χ^2 test is significant ($p < 0.01$) and shows a double periodicity of 54.5 hours and 7 days with magnitude variations of 0.5 mag ± 0.1. (See Fig. 1 where each observer is represented by a different symbol, and the average period of the variations is 54.5 hours ± 8 hrs decreasing in amplitude over a 7-day period).

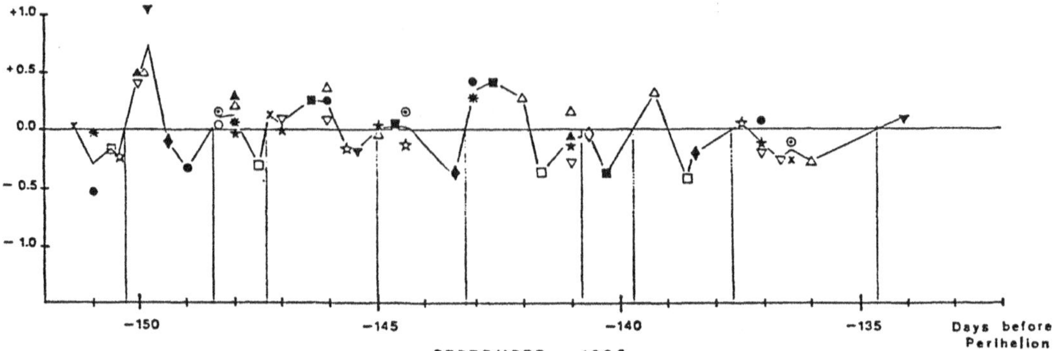

References

1. Festou, M. et al.: "Periodicities in the light-curve of P/Halley and the rotation of its nucleus", *Astron. and Astrophys.*, in press
2. Edberg, S.J. et al.: "Observational factors affecting studies of comet P/Halley's visual light-curve", Proc. 20th ESLAB Symposion on the Exploration of Halley's Comet, Heidelberg, 1986 Oct. 27-31

I thank Dr A.C. Levasseur-Regourd and J.C. Merlin for their comments on this paper.

P/Halley: The Disconnection Event of 1986 April 11

P.M. Bergé[1], *G. Mahoux*[2] and *A.C. Levasseur-Regourd*[3]
[1]Residence "La Dauniere" Bat. D, F-91940 Les Ulis, France
[2]12 rue du Dr. Kurzenne, F-78350 Jouy-en-Josas, France
[3]Université Paris VI, Service d'Aeronomie, F-91370 Verrières-le-Buisson, France

This work was carried out by an expedition organized by the S.A.F. in 1986 April to La Réunion. The aim was to observe and photograph P/Halley, within the framework of IHW, as part of the Island Network in the southern hemisphere. To be more precise, our work consisted of studying large-scale phenomena: the structure, dynamics and possible disconnection events in the plasma tail. We were lucky enough to observe one of the latter on the night of April 11/12, and describe it here.

For the Island Network, IHW had a number of Schmidt telescopes (Celestron 8). One was lent to the S.A.F. and this is what we used. This telescope has a focal ratio of 1.5, with a 200-mm (8-inch) objective and 300-mm focal length. We used only Kodak TP 2415 film, hypersensitized in forming gas (24 h at 60°C).

We obtained three photographs of the discontinuity, only the last two being reproduced here (as negatives). The guiding was on the comet's central condensation. In addition, 3 other photographs were available, taken by Gerhart Klaus in the Namib Desert, using a Schmidt similar to ours (but outside the IHW framework).

Photo. No.	Site	Date (UT)	Exposure	Filter
1	La Réunion	Apr.11.868	9 m	none
2	"	Apr.11.881	20 m	"
3	"	Apr.11.928	60 m	W47B
4	Namibia	Apr.11.889	10 m	none
5	"	Apr.11.920	10 m	"
6	"	Apr.12.090	10 m	"

On photo 2, the coma, the rather wide dust tail, and straighter plasma tail, which shows a very distinct disconnection near the coma, can all be seen. On photo 3, the star trails are much longer, the exposure being 60 minutes. The blue W47B filter has practically extinguished the yellow-coloured dust tail, but retained the bluish plasma tail. In the old tail, emitted before the disconnection, parallel structures can be seen, which suggest the existence of turbulence in this part of the plasma tail. The last photograph taken by G. Klaus about 4 hours after out third photograph (and not reproduced here), shows considerable evolution in the plasma tail, near the disconnection.

Reduction of the photographs required the calculation for each of them of the equatorial coordinates of the comet, the Δ and R distances from the Earth and the Sun, the antisolar direction, and the phase angle (the Earth-comet-Sun angle, necessary for the measurement of real distances along the plasma tail). We used the photographic atlas by Papadopoulos and Scovil to determine the position of the cometary nucleus on each plate, as well as the direction of North and the antisolar direction, which allowed us to measure the tail's directions before and after the disconnection. We measured the speed of recession of two specific points on the plasma tail: the one where the old tail became disconnected, and the one where the new tail reformed. We found 50 ± 5 and 80 ± 20 km/s respectively, which agree with the values usually found for the recession velocities of cometary tails. Despite the error in the measurements, caused by the difficulty in identifying exactly the same points on the differnet plates, the velocity of the new tail seems significantly greater than that of the older one.

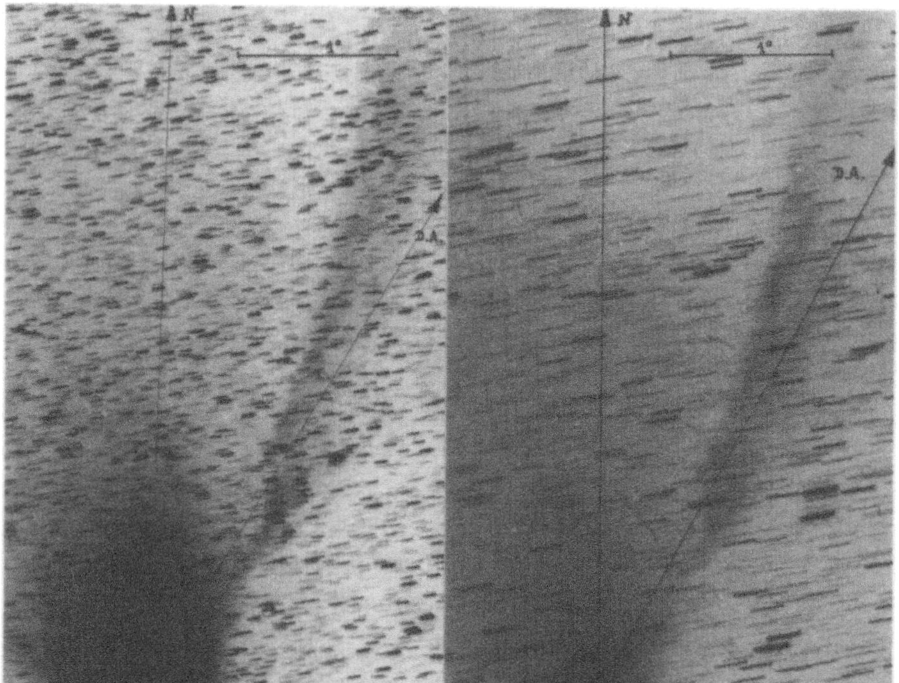

Photos 2 (left) and 3 (right)

The interpretation currently given for disconnection events, proposed by Malcolm B. Niedner and John C. Brandt (1), calls for the comet to pass through a region where the magnetic field carried by the solar wind changes sense. It is known that the ions of cometary origin that form the plasma tail are channeled along the lines of magnetic force. In a region where the field rapidly changes its sense, the plasma tail is dislocated, and a new tail forms slightly later, when the new magnetic field has stabilized. However, as far as the event that we are considering here is concerned, Niedner himself has cast grave doubt on this explanation (2). On April 11, the comet would not have crossed a boundary between magnetic-field sectors of opposite polarity. However, the interplanetary probes IMP-8, Pioneer Venus Orbiter and ICE have revealed the existence of small-scale reversals of polarity outside the neutral sheet, where such reversals are normally expected. This could be the cause of the April 11 event, which would therefore be a sign of local turbulence in the solar wind and its magnetic field.

References

1. M.B. Niedner, J.C. Brandt: Astrophys. J., **223**, 655 (1978)
2. M.B. Niedner: COSPAR Meeting, Toulouse, France, 1986 July

A Study of Disconnection Velocities in the Plasma Tail of P/Halley

Roland Caron[1], *Serge Koutchmy*[2], and *Michel Sarrazin*[1]
[1]19 av. Marcelin-Berthelot, F-92390 La Garenne, France
[2]Institut d'Astrophysique de Paris, 98 bis Bd. Arago, F-75014 Paris, France

During a photographic campaign on Comet P/Halley in 1986 April, a considerable number of photographs were obtained, especially on April 8, of a disconnection event in the plasma tail. Initial details were presented at the Symposium in Heidelberg at the end of 1986.

The photographs were obtained with a camera five examples of which were specially constructed (by Carron), and known as the K600. The objective is a 100-mm OG, focal length 600 mm and the image is corrected with a field lens. The camera body is a Hasselblad, taking 120 roll film, which was Agfa 1000.

Each camera was mounted on a guide telescope with means of sidereal and cometary tracking. The sites used for the expeditions were New Caledonia, La Réunion, Namibia, Chile and Tahiti. At Tahiti, a second instrument consisting of a 135-mm objective in front of an image-intensifier was used to obtain other images used in this study.

Images selected for use were: La Réunion 4; Namibia 1; Chile 9; and Tahiti 8. It should be noted that measurements were made on prints, the scale of which was rigorously controled to be the same for each plate.

Graphic software was developed to give the maximum precision for all the positional elements of the comet within the star-field and of the position of the plasma tail. The position of the cometary nucleus is calculated from the location of several known reference stars. The anti-solar direction is determined and drawn in, graduated in 10^5 km as a function of phase angle, i.e. in the tail's real position with respect to the line of sight.

Various other values were calculated for the time of each exposure. The screen image was dumped to a printer and copied onto transparent overlays. Alignment of the transparencies was carried out by carefully superimposing the appropriate stars (with due regard to the trails resulting from following the comet during the 20–30 minute exposures used for the K600 images). This procedure allows all the measurements of displacements within the plasma tail to be related to the same reference point (that of the nucleus according to published ephemerides).

The measurements obtained as outlined above are shown in the Figure. As will be seen, the mass of ionized gas representing the disconnection event investigated in not carried along by the solar wind with either a uniformly accelerated or decelerated velocity. The first two accelerations take the velocity of the knot of gas to 57 km/s and 52 km/s, while the minima are of the order of 41 km/s. The end of the curve shows a considerable increase in velocity. This is derived from the last two images obtained of this disconnection.

This study shows that particularly violent events occur in the gas tails of comets like P/Halley, driven by sudden changes in the velocity and direction of the solar wind. It should be noted that there is no evidence for tangential displacement – that is in the plane contained by the image plane.

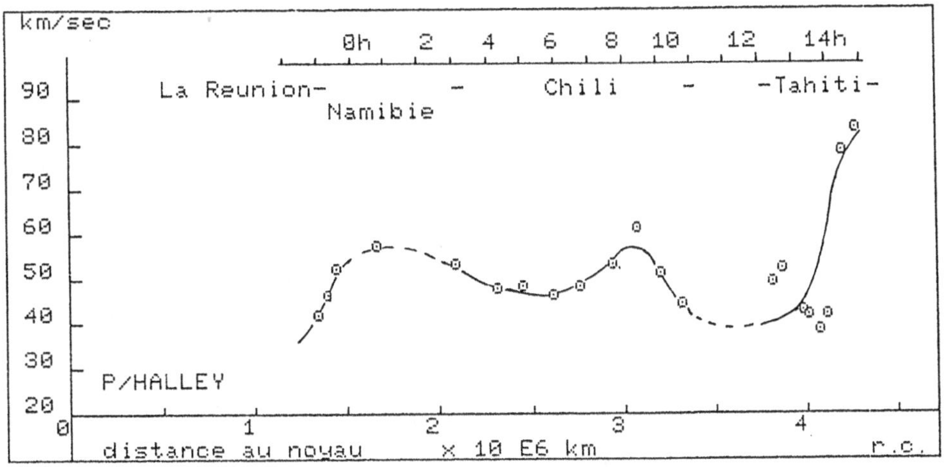

Velocity curve for the 1986 April 8 disconnection event

Reference

R. Caron et al.: 20th ESLAB Symposium on the Exploration of Halley's Comet, Heidelberg, 1986

Analysis of the Light-Curve of Comet P/Halley in 1985–1986

David Valls-Gabaud[1,2] and *Angel Gómez*[1]
[1] Agrupación Astronómica de Madrid, Apartado 1039, E-28080 Madrid Spain
[2] Institut d'Astrophysique de Paris 98 bis, Bld. Arago, F-75014 Paris, France

Summary. We analize the light curve of comet P/Halley resulting from 187 observations made by amateurs. We try to derive some constraints on the physical properties of the comet through the application of the semi-empirical photometric model (SEPM) developped by R.L. Newburn.

The m_1 (Fig. 2) and $V(1,0)$ light curves show a clear transition phase in the pre-perihelion apparition, between $\sim$ 1.9 and $\sim$ 1.2 A.U. which supports the division into 3 parts made by Hasegawa. It is remarkable to find that the light curve obtained by amateurs is so close to those obtained by experienced observers.

We derive an effective equivalent radius of $\beta = 4.9 \pm 0.8$ (1σ) km, and a geometrical albedo of $p_v = 0.06 \pm 0.02$ (1σ) using the method suggested by Delsemme and Rud.

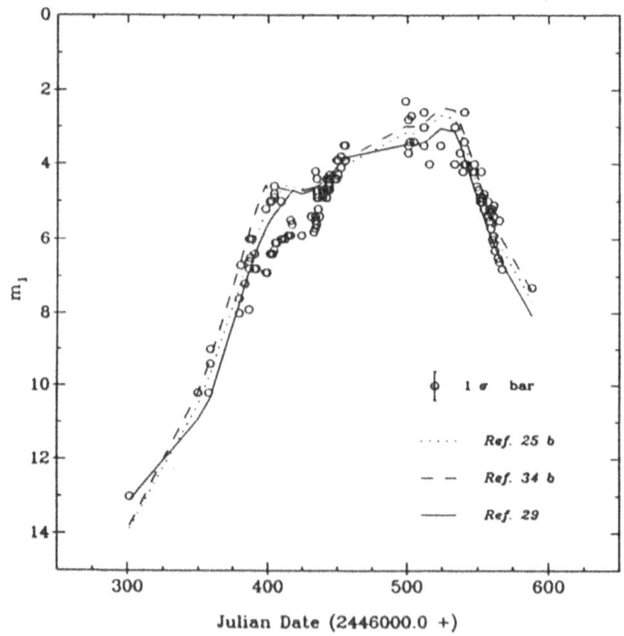

Fig. 2. A.A.M. light curve of comet P/Halley in 1985–1986

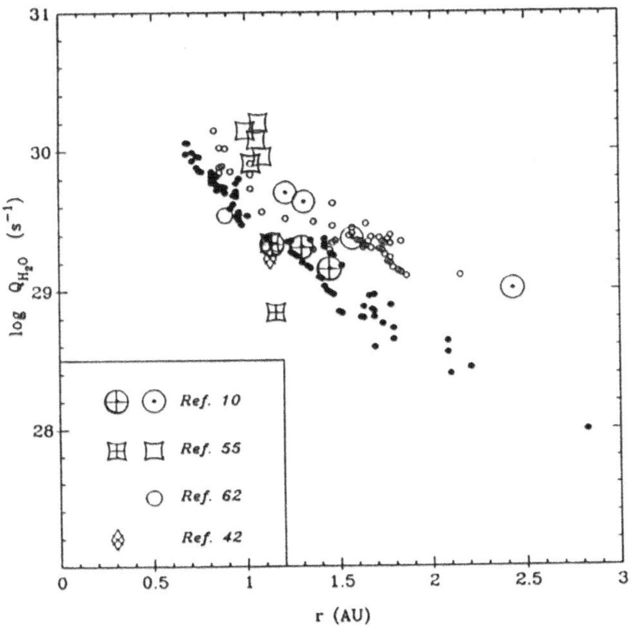

Fig. 5. Water production rate as a function of heliocentric distance. Post-perihelion (*open symbols*), pre-perihelion (*filled symbols*)

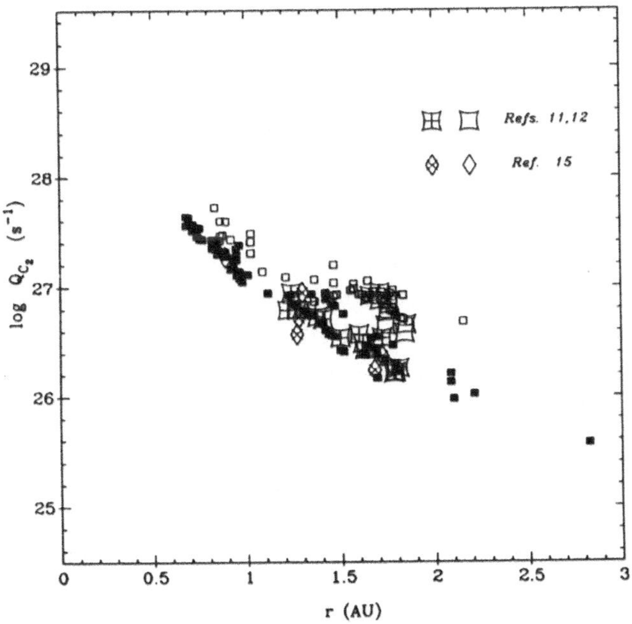

Fig. 6b. C_2 production rate, assuming a linear ($m = 1$) dependence on Q_H. Symbols as in Fig. 5

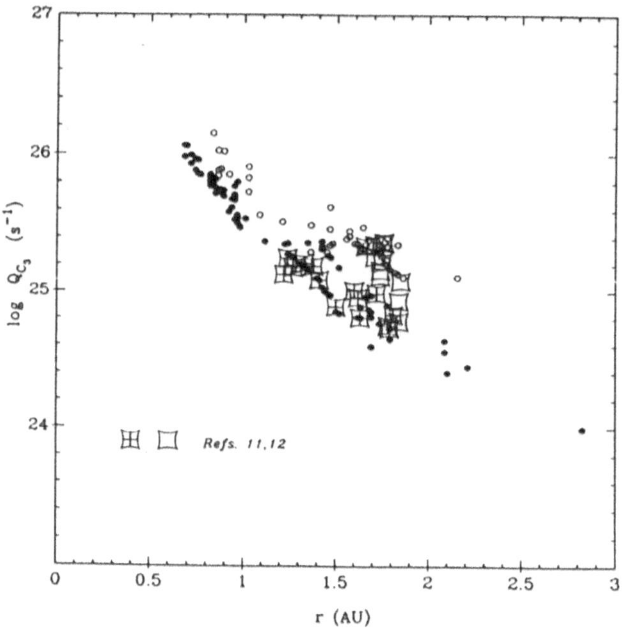

Fig. 7 C_3 production rate. Symbols as in Fig. 5

The water production rate, obtained from a straightforward application of the SEPM, is illustrated in Fig. 5, where it is compared to different spectrophotometric and narrow-band photometry measures found in the literature. Note that the strongest pre-/post-perihelion asymmetry (a factor of 4–6) takes place for heliocentric distances larger than about 1.7 A.U.. This asymmetry is found in each production rate computed. A tentative explanation may be the rotation associated with a rapid change of declination of the sub–solar point, leading to the sudden sublimation of underexposed nuclear regions, as suggested by P.R. Weissman.

The C_2 production rate is shown in Fig. 6b, where the best results are obtained with a *linear* dependence on the H rate, and a mixing constant of 1.5 10^{-32}, consistent with the C_3/CN ratio. The derived relative compositions are found consistent with other estimations.

We think the SEPM provides a powerful tool to find correlations to be used as constraints to the physical modelling of comets, self-consistently within factors of 2–3. This kind of analysis should be useful for amateurs that wish to interpret their own observations, provided that a detailed light curve is obtained.

A detailed account of this work has been submitted for publication to the *Journal* of the British Astronomical Association, and may be obtained from the authors upon request.

New Trends in the Discovery of Comets by Amateurs

J.C. Merlin
17 rue P. Mendès-France, F-71200 Le Creusot, France

Abstract

The ways in which amateurs may increase their chances of discovering comets were described. Amateurs discover a considerable proportion of comets, but their contribution is greater among long-period comets. The relative proportions of short-period comets is revealing: 1 in 2 for professional astronomers, 1 in 8 for amateurs. Professional discoveries are largely a by-product of searches for asteroids: the comets have low inclinations ($i < 30°$) and are faint ($m_1 > 14$). Amateurs are most successful in areas close ($< 60°$) to the Sun. Typical apertures are 150-mm, and most comets are brighter than magnitude 10.

Past results show that there is a strong correlation between number of observers and number of discoveries. Several amateurs have discovered comets at great elongations with larger telescopes (≥ 400 mm). The comets tend to be fainter ($10 < m_1 < 12$). Several comets have been discovered with simple equipment (200- or 300-mm telephoto lenses) down to magnitude 13.

Calculations of the distribution of discoverable comets show that an average of 14 comets ($9 < m_1 < 14$) are missed per year.

References

Kresak, L., 1982: "Comet discoveries, statistics, and observational selection" in *Comets*, ed. Wilkening, L.L., 56–82

Everhart, E., 1967: "Intrinsic Distribution of Cometary Parhelia and Magnitudes", Astron. J., **72** (8), 1002–11

[*See also* the invited paper by B. Marsden – Eds.]

Amateur Astronomers and the Recovery of Periodic Comets

Patrick Martinez
10 rue Alphonse-Daudet, F-31200 Toulouse, France

Abstract

Most periodic comets are recovered by professional astronomers at faint magnitudes (18–21). Only Tsutomu Seki, using a 60-cm telescope has been an exception to this rule. Amateurs having access to large telescopes (such as the T 60 at Pic du Midi) now stand a chance of success. For example, using the T 60, the author photographed P/Halley at mag. 19–20 on forming-gas hypered Kodak TP 2415 in a 1-hour exposure. Experiments by C. Buil suggest that the same telescope can reach magnitudes 21–22 with a CCD array cooled to −50°C and a 30-minute integration time.

References

Martinez, P.: "Première détection europée̊ne amateur de la comète de Halley", *L'Astronomie*, 1985 June, p.281
Martinez, P.: "L'utilisation du T 60 pour la redécouverte des comètes périodiques", *L'Astronomie*, 1987 May, p.247
Merlin, J.C.: "La redécouverte des comètes périodiques" in *Astronomie – le Guide de l'Observateur*, Paris, 1987, pp.417–20
[*See also* the paper by C. Buil on the use of CCDs – Eds.]

Shape and Frequency of Outbursts of P/Schwassmann-Wachmann 1

J.C. Merlin, D. Fayard, and *E. Gérard*
17 rue P. Mendès-France, F-71200 Le Creusot, France

Abstract

Observations made since the comet's discovery (1927) were analyzed and reduced to remove observational bias. Comparison was made with possible mechanical effects on the nucleus (tidal forces, impacts). The observations appear to best fit an internal mechanism (exothermic chemical reactions) as proposed by Whipple.

[The text of this paper was received, but was of such extreme length that it could not be included in full, as the author requested. – Eds.]

Astrometry of Comet Halley

Hadi Hadavi Biruni Observatory, Shiraz University, Shiraz 71454, Iran

[This contribution was delivered, but neither text nor abstract have been received. – Eds.]

Observation and Analysis of Comet Tails

R.L.W. van de Weg
von Hemskerkstraat 29, NL-7622 JG Borne, Netherlands

Abstract

Details of H_2O and dust production in P/Halley are mainly derived from amateur observations of cometary tails, leading to a gas/dust mass-ratio. The rotational period (2.21 day) is also found from amateur observations. Velocity and acceleration of the plasma and velocity and direction of the solar wind are also derived, together with size and indications of the mass-distribution function for dust particles.

These techniques have been applied to other comets, giving interesting results on differences between various comets. There is a lot of information available in cometary photographs. The coordination and archival storage of such information is urgently required.

[This paper was delivered, but no text was received. – Eds.]

Photometric Analysis of the Solar Corona on 1984 November 22/23

Roland Caron[1], *Christian Nitscheim*[2], and *Michel Sarrazin*[1]
[1]Société Astronomique de France, 3, rue Beethoven, F-75016 Paris, France
[2]Institut d'Astrophysique de Paris, 98 bis, Bd. Arago, F-75014 Paris, France

Introduction

The total solar eclipse of 1984 November 22/23 was observed from a French naval vessel off Noumea (New Caledonia) in the Coral Sea. The observations were made by a team from the S.A.F., led by Serge Koutchmy and Christian Nitschelm of the Institut d'Astrophysique de Paris. The duration of the eclipse was considerably longer off New Caledonia (1 m 39 s) than in Papua-New-Guinea (55 s). However, the observational site – on board a ship well out to sea – greatly restricted the use of long-focal-length instruments that are normally employed. The equipment used to photograph this eclipse therefore was chosen with motion (pitching, rolling, etc.) of the support in mind. Short focal lengths were preferred, although one refractor with a focal length of 1.5 m was used as a trial.

Observations and Results

The geographic location of the observing site was 165° 11′ 36″ East, 20° 03′ 36″ South. The ship was practically on the central line. The altitude of the Sun was 50.5° and its azimuth 94°.

The results obtained were better than expected. In particular the 1.5-m f/15 refractor, fitted with a Zenith 80 body, obtained two images that did not show any motion. The film used was Ektachrome 400, 6 × 6, a good compromise between fine-grain and the speed needed when the support is in motion. The other objectives used either 35-mm Ektachrome or much faster films, such as 35-mm Fujicolor 1600.

With a radial neutral density filter, we processed the high-resolution images in order to accentuate fine detail in the corona. The resulting image showed coronal structure at the time of the eclipse, and jets and the magnetic-field lines could be seen easily.

Photometric Analysis

Using the best images and video-processing equipment at the Institut d'Astrophysique de Paris, we were able to construct sets of isophotes. Calibration was carried out by using existing coronal models. These isophotes allow the ellipticity of the corona to be calculated. This is given by the following equation:

$$\varepsilon = \frac{R_{eq}}{R_{pol}} - 1 \quad .$$

Near the solar limb, this ellipticity follows a linear law:

$$\varepsilon = a + b\left(\frac{R_{eq}}{R_\odot} - 1\right)$$

where a and b must be determined for each eclipse.

The curve of the change in the ellipticity factor enables us to determine the values of $a - b$ and $a + b$ as -0.289 and 0.381, which gives values of 0.046 and 0.335 for a and b. These values appear high, especially $a + b$ (ε at $2\,R_\odot$), given the period of the cycle (beginning of solar minimum, $\phi = -0.23$).

Conclusion

This study shows that it is perfectly possible to observe a total solar eclipse at sea and obtain results that can be of some value. This work is only a beginning, however, and a more complete study is required.

The corona was near its least active phase (solar minimum in 1986), but it still showed distinct jets and polar plumes.

Even with the limited equipment and somewhat precarious observing condition, interesting results were obtained, confirming that observation of total solar eclipses by amateurs advised or trained by professionals should be standard practice for future eclipses.

References

Koutchmy, S. et al.: Astron. Astrophys., **69**, 35
Koutchmy, S., Nitschelm, C., 1984: Astron. Astrophys., **138**, 161–3
Koutchmy, S., Stellmacher, G., 1976: Solar Phys., **49**, 253
Nitschelm, C., Koutchmy, S., 1984: L'Astronomie, **98**, 393–9
Nitschelm, C., Sarrazin, M., 1987: L'Astronomie, **101**, 207–211; 1984, L'Astronomie, **98**, 393–9

Chromospheric Eruptions Observed with a Wide Slit

François Costard
22 av. Perichont, F-75016 Paris, France

Abstract

The method of placing the solar limb tangential to the slit, which is then widened, has been employed. The following phenomena have been observed: fast radial motion of low loops; extension of the main loop into the corona; ejection of material from the top of the loop; rotation of the structure as it relaxes.

Reference

Costard, F.: "L'eruption chromosphérique du 12/09/85", *L'Astronomie*, 1987 Jan., p.21–7

Amateur/Professional Cooperation in 2 Solar Studies

T. Roudier, R. Muller, J.C. Hulot, and *F.Vaissière*
Observatoire du Pic du Midi et de Toulouse, 14 av. Edouard Belin,
F-31400 Toulouse, France

Abstract

The modification of properties of granules around magnetic flux tubes has been studied for the first time from photographs (at $\lambda = 4803$ Å and 5770 Å) taken with the 50-cm refractor at Pic du Midi. Statistically, these granules are more numerous, smaller, and more elongated than other granules. During their first two minutes of life they show very pronounced radial orientation to the magnetic flux tube.

Angular measurements on the same granules have a precision of $\pm 10°$, which is sufficient as theoretical studies show that they rotate by 360° in the course of their life. Initial results appeared to show that only explosive granules had intrinsic rotation, but further examination showed that it is a general trend. It seems that the granules do rotate significantly, but that there is a more important "push-pull" effect, in agreement with A. Title's theory drawn from SOUP images.

[Submitted to *Solar Physics*]

The IOTA Expedition Results from the Total-Annular Eclipse in Gabon, 1987

Paul D. Maley
15807 Brookvilla, Houston, TX 77059, U.S.A.

Abstract

The results of the IOTA expedition to Gabon to observe the rare total-annular eclipse of 1987 March 29 were described. Multiple stations were used to determine data from Baily's Beads about the contact times. The techniques used and their application to future eclipses were evaluated. An additional aim was to develop a database to resolve corrections to the solar radius at specific epochs in order to determine the amplitude of time-dependent oscillations of the Sun's surface.

[The text was not received. – Eds.]

Meteor Studies

B.A. Lindblad
Box 43, S-221 00 Lund Sweden

Historically meteor astronomy is one area where amateurs have always been able to make significant contributions. In fact, in the 19th century, it was amateur naked eye and telescopic observations which laid down much of the foundations of meteor astronomy. References to this work can be found in any textbook on meteors. The 19th century observers concentrated on counting meteors, estimating magnitudes and plotting the meteor paths on star maps. Their main interest was to determine hourly rates and shower radiants. An important milestone was Denning's radiant catalogue (Denning 1882), which included 4367 shower radiants. Although it is now believed that many of these radiants are spurious, the catalogue is still a useful reference. Unfortunately Denning and other 19th century observers often combined sporadic meteors observed on different nights into a minor stream radiant. This habit of "radiant hunting" is even today quite popular among some amateur observers. However, in all fairness it should be emphasized that most of the 20th century amateur meteor observers applied very strict criteria to their radiant determinations. Names such as J.M. Prentice in Great Britain, R.A. McIntosh in New Zealand and R. Rigollet in France may be mentioned.

The extensive development of professional meteor programs after the second World War involving photographic, radar and image orthicon techniques added vastly to our knowledge of meteors and thereby no doubt caused a temporary decline in amateur activities. Fortunately there is today a new generation of amateur observers who work very professionally. This new generation observes meteors by classical techniques, but in addition it uses electronic devices such as tape recorders, personal computers and forward scatter radio systems, thereby successfully competing with the professional scientists. Today there is a high degree of cooperation between the amateur observers. A number of amateur meteor societies and their publications are listed in Table 1. The listed societies have existed for a very long time - in one case for more than 50 years. The Dutch and Belgian reports are referenced in Astronomy and Astrophysics Abstracts.

In what way can today's amateur observers contribute to meteor astronomy? The answer is obvious: The professional astronomer has neither time nor money to maintain a constant sky watch. In the 1950's and 1960's a number of meteor radars kept a

Table 1. Meteor societies and reports

Country	Name	Publications
Australia	NAPO Meteor Section	Various reports
Belgium	Vereniging voor Sterrenkunde Werkgroep Meteoren	Werkgroep nieuws - meteoren (WGN)
DDR	Arbeits Kreise Meteore	Mitteilungen der A.K. Meteore
Great Britain	Meteor Section of the BAA	Newsletter BAA Meteor Section
Japan	Nippon Meteor Soc.	NMS Astron. Circ.
Netherlands	NVWS Meteor Section	Radiant - J. Dutch Meteor Soc.
USA	American Meteor Soc.	Various reports
	Callahan Meteor Soc.	Meteor News (Jacksonville, Florida)

semi-permanent watch on meteor activity. Unfortunately, most of these programs were terminated in the 1960's and 1970's. The demise of the meteor radars in most western countries means that any new meteor shower which suddenly appears in the night sky will pass undetected, unless it is detected by an amateur group.

In this talk I will highlight three cases where amateur observers recently have made significant contributions to meteor astronomy.

1. Observations of the Sigma Puppid stream associated with comet Grigg-Skjellerup

In 1964 P/Grigg-Skjellerup made a close approach to Jupiter. The Jupiter encounter perturbed the comet's orbit, so that from 1967 and on very close approaches to the Earth's orbit occur. Sitarski (1964) predicted that a meteor shower would be observed on April 23 in 1967, 1972 and 1977. The theoretical radiant of this stream is $\alpha = 109^{o}$, $\delta = -45^{o}$. The far southern declination of the radiant implies that only a limited number of amateur groups are available for observing this meteor shower. Since the stream is newly born, total shower duration may be only a few hours. Thus the shower may be observed only over a limited part of the Earth. In 1967 the comet-Earth distance at the nodal point was only 0.0027 a.u., but the Earth crossed the nodal line 97 days later than the comet. It was doubtful if fresh cometary debris had spread that far along the orbit.

No Sigma Puppid shower was observed in 1967 and 1972 indicating that the cometary debris was still in close proximity of the comet. In 1977 an intense, shortlived Sigma Puppid shower was observed by Australian amateurs (Wood 1986). Continued visual observations of this shower in 1982-84 by the same group indicated that the cometary material was gradually spreading along the orbit and in the orbital plane of the comet. For, perhaps, the first time we can study the birth and subsequent evolution of a meteoroid stream. The Australian amateur observations are our only source of information on this unique meteor stream. A review of these observations is given in Lindblad (1986a).

2. An amateur list of meteor orbits

A Japanese amateur, M. Koseki, has compiled a list of all photographic two station meteor orbits published in the scientific literature and searched this sample on a micro-computer for meteor streams (Koseki 1986). Koseki has also checked the original orbits and found a substantial number of errors in the published orbital elements! Koseki has kindly supplied the IAU meteor data center in Lund with a list of these errors. This is a very impressive piece of work done by the new computer generation of amateur astronomers.

3. Observations of the Giacobinid meteor stream by forward scatter techniques

Forward scatter radio techniques for the study of meteors is gaining increasing popularity amongst amateur groups. The technique is in principle quite simple - you listen to the carrier frequency of a TV station at some 1000 kms distance and record meteor burst signals on a tape recorder. However, the geometry of meteor forward scatter is quite complicated, and interpretation of the data demands considerble skill. For a discussion of the observability function of forward meteor scatter see Steyaert (1987) where additional references can be found.

During the 1985 recurrence of the Giacobinid meteor shower (associated with comet Giacobini-Zinner) several amateurs made radio recordings of the intense, short duration Giacobinid display. A British amateur Mason (1986) has studied the hourly meteor burst rate versus time. Mason's hourly rate curve agrees closely with a curve obtained by Lindblad (1986b) using the Onsala meteor radar. The time of peak activity agreed to within 1 min.

References

Koseki, M., 1986, Brit. Astron. Assoc., 96, pp. 232-240.
Lindblad, B.A., 1986a, ESA SP-250, vol. III, pp. 399-400.
Lindblad, B.A., 1986b, ESA SP-250, vol. II, pp. 229-231.
Mason, J., 1986, ESA SP-250, vol. III, pp. 379-384.
Sitarski, G., 1964, Acta Astronomica, 14, pp. 25-37.
Steyaert, C., 1987, Werkgroep nieuws - Meteoren, 15, pp. 90-93.
Wood, J., 1986, priv. comm. See also Western Australian Meteor Soc. Bull., 80.

The "Sonate" Campaign

Jean Mazel[1] and *Robert Futaully*[2]
[1]12 rue Montmorency, F-34500 Béziers, France
[2]Observatoire du Pic du Midi et de Toulouse, 14 av. Edouard Belin, F-31400 Toulouse, France

"Sonate" (Sondages acoustiques pour l'évaluation de la qualité d'images des télé-scopes) – Acoustic sounding to estimate the quality of telescope images – is a prime example of professional/amateur collaboration as it involved 3 professionals and 22 amateurs, who, in order to collect the maximum amount of data, relayed one another in 10-day shifts over a period of 4 months in 1984, using the 1-m telescope at the Pic du Midi. A preliminary campaign with the 2-m telescope in 1982 had shown that the profile of sound waves reflected from the layers above a site could be correlated with the quality of telescopic images at the same site.

Each team collected two series of data:
1) Beginning two hours before nightfall, after reading temperature, atmosphere pressure and humidity, a series of SODAR measurements was taken. SODAR, mounted on a building near the 1-m, uses a parabolic dish pointing to the zenith to emits a series of 2000–Hz blips. The echoes from heights of 800 m above the site are received by the same dish. This layer of the atmosphere contains about 80% of the sources of turbulence affecting the site, so the information is capable of providing an adequate model of the atmospheric structure through which the telescopes will have to work. Integration of profiles obtained at 5-second intervals allows an average value of R_o, a parameter expressing the variations in the refractive index of the air, to be calculated.
2) After nightfall, objective measurements of R_o were made by the SYLVAIN equipment (a camera giving a real-time stellar image, digitized and integrated to calculate R_o. The measuring procedure could take 20–30 minutes, but was helped by a detailed check-list. The objective value of R_o obtained by SYLVAIN was obviously compared immediately with the SODAR value.

Of the 108 days of the campaign, 60 nights' values could be used. The correlation between the SODAR and SYLVAIN values was 0.895. The rare disagreements could explained. In general a SODAR prediction is always reliable when it indicates poor resolution. It is 90% reliable when it indicates good, or excellent, resolution. The results were all that were desired, showing that SODAR is a useful tool in predicting image quality, and can be used for site testing, and at existing sites. SODAR is now permanently installed at the Canada-France-Hawaii Telescope site and at ESO in Chile.

What did amateurs get out of it, because the campaign was conceived and arranged by professionals, and the amateurs were just the labour force. There were compensations. Not only was there the chance to work at the Pic du Midi, but there was also the incentive of having the use of the 1-m telescope for their own work after the SODAR and SYLVAIN measurements had been made.

The first teams did find reluctant acceptance by some of the professionals at the Pic, who seemed to resent amateurs being present in a professional establishment and feel that they were using the 1-m under false pretences. Relationships changed very considerably, however, over the 4 months. By the end of the campaign, SODAR was part of the fixtures at the Pic, and at the evening meal the R_o prediction was a topic of general conversation.

Although the first teams may have encountered a rather strained atmosphere, the later ones were enthusiastic about their participation. Having conducted a project, envisaged by one set of people, and carried out by another, we have learnt quite a lot. I think this applies to effective future collaboration between amateurs and professionals. Both must have something to gain.

Two Solar Observation Programmes

Marie J. Marthes, Jean M. Malherbe, and *J. Vial*
Observatoire de Paris, 5, pl. Jules Janssen, F-92195 Meudon Principal Cedex, France

The Solar Section of the Societé Astronomique de France has organized various programmes designed to enable amateurs to carry out high-quality and scientifically useful observations. Two particular programmed will be described.

The Study of the Proper Motions of Sunspots

This was started in 1974, and was aimed at all potential observers. It was carried out under professional guidance and was designed to determine the speed of rotation of major spots and their proper motion in heliographic latitude and longitude. Further statistical study aimed at examining the data as a function of the period of the cycle (the latitude effect); relative positions (interactions between groups); departures from average motion (mean differential rotation). Sunspots are tracers of the magnetic field and there appearance is linked with activity in deep layers of the Sun that are not directly accessible to observation. The proper motion of spots in their first days of life is determined by these inner layers, so these studies are relevant to understanding convection within the Sun.

Observing was in white light and the programme covered the period from 1973 to 1982 (Cycle 21). A minimum of standardization was required, observers being able to observe whenever they could and would, with their own instruments and techniques. They were encouraged to continue to derive Wolf numbers themselves and, finally, were only given instructions and advice when they felt they wanted them. In return for their efforts they learnt to determine the position of the solar axis, to follow solar rotation from the motion of sunspots, to calculate heliographic coordinates, and possibly to take part in the analysis of the data received.

This programme, covering the rise to maximum of Cycle 21 and the subsequent decline was a great success. More than 40 000 positional determinations were received, exceeding all expectations, and causing some problems for the Section in handling all the information. Although theoretically finishing with Cycle 21, this programme may be revived as spots become more numerous (the maximum of Cycle 22 is predicted to be 1990–1).

Solar Prominence Programme

The programme is due to start in 1988 and involves systematic H-α observations of prominences at the solar limb, with particular reference to two types of instability: those associated with quiescent prominences – sudden disappearances (which remain poorly understood), and bridges by which material passes from one prominence to another – and those events not linked to existing prominences, such as surges, ejection of material and eruptions. In the latter case the phenomena are associated with active centres, plages, and spots.

Three types of study are proposed: morphology, kinematics – for example, the various parameters of the apparent motion against the plane of the sky can be calculated and compared with typical values of velocities, time, and acceleration – and finally, comparison of the phenomena as seen against the sky with similar features seen against the solar disk, and thus trying to understand the three-dimensional structure.

Analysis will be by a combined team of professionals and amateurs. Far from being a hindrance, the great variety of structures that can be observed are an added incentive to obtain as many observations as possible so that they may be classified and brought into an overall scheme. The wide variety of equipment used and the geographical distribution of amateurs will be of great value in carrying out this work.

Long-term Monitoring of Solar Activity by Amateur Astronomers

Klaus Reinsch
Wilhelm-Foerster Sternwartem Munsterdamm 90, D-1000, Berlin 41, Germany

Introduction

Professional solar astronomy concentrates on the study of the atmosphere and interior of the Sun. Little attention is given to "classical" programmes, mainly statistical investigations of solar activity. Although the main properties of phenomena associated with the solar cycle seem to be understood there are still enough details to be explained, making it worthwhile monitoring different indicators of solar activity, even if no immediate results are to be expected. Such routine observations are ideal work of amateur astronomers.

Members of West German local astronomical societies founded the journal *Sonne* in 1977 to combine their efforts on solar observations. The first issue was presented at a conference on amateur solar observation held in Berlin in April 1977. *Sonne* is compiled by an editorial staff of 23 amateurs from all over West Germany, and is distributed among nearly 500 readers in 20 countries. With the increasing number of foreign readers, the main articles in *Sonne* are provided with English abstracts.

Common efforts of 28 amateurs led to another remarkable product: in 1983 the "Handbook for Solar Observers" was completed after 5 years work. In 700 pages all topics of current amateur solar observation are discussed: white light, Hα, radio, and solar eclipses, including all relevant techniques and references. An English translation is in preparation.

Observing Programmes

The staff of *Sonne* also coordinate different observing programmes:
Sunspot numbers; new sunspot indices; daily charts of the Sun; sunspot positions; white-light faculae; solar photography; Hα observations; light bridges; naked-eye sunspot observations; solar eclipses.

The sunspot network currently has more than 130 observers worldwide. Daily Wolf sunspot numbers are evaluated from about 17 000 individual observation per year. The basis for the reduction of the sunspot numbers are the data obtained by a group of experienced and qualified observers who are selected yearly. Definitive sunspot numbers now cover nearly a complete cycle (see Fig. 1). Smoothing of the monthly means is done by an improved method using a polynomial weighting function over 17 months (instead of the traditional box-like weighting function).

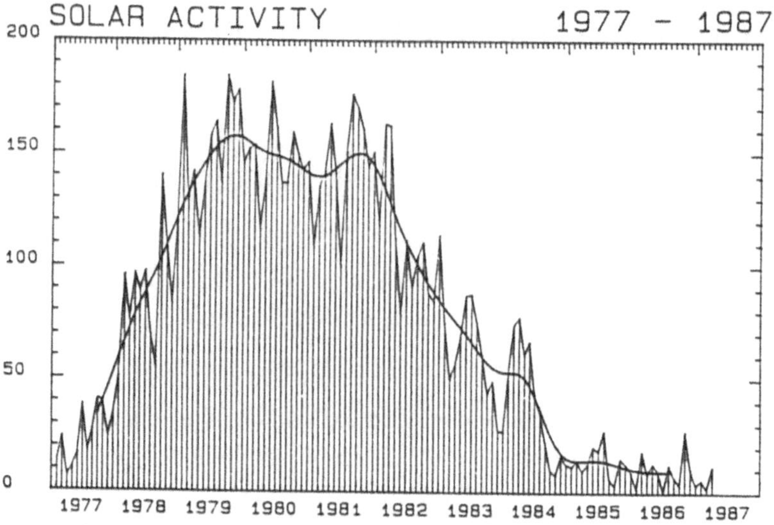

Fig. 1. Monthly and smoothed means of the definitive sunspot numbers evaluated by the *Sonne* network

The comparison between the SIDC and the *Sonne* sunspot numbers is shown in Fig. 2. Although there is a good match overall, the *Sonne* sunspot numbers tend to lie above the SIDC values at high activity and below them at low activity. But a similar relation is noted when comparing the SIDC data with the sunspot numbers obtained by the Zurich observatory, which was been the international standard until 1980.

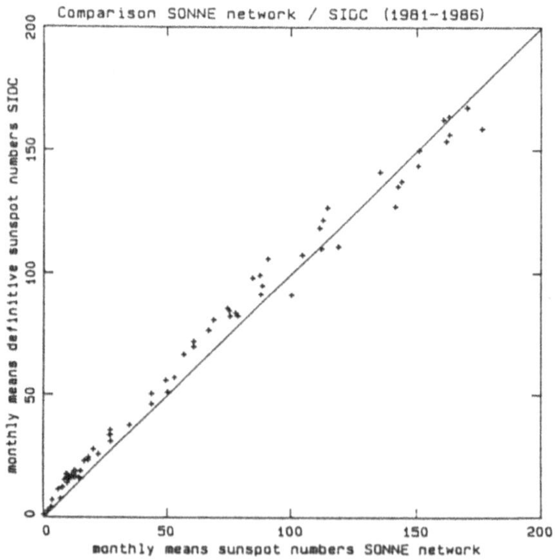

Fig. 2. Comparison between SIDC and *Sonne* sunspot numbers (1981–1986)

It is well known that the sunspot number defined by Wolf is not the best measure of solar activity, because small and large sunspot groups are given the same weighting factor of 10. A new sunspot number that uses different weighting factors according to the Waldmeier classification of a group, has been introduced by R. Beck. This gives an estimate of the sunspot area:

$$\mathrm{Re} = \sigma_i \, G_i \, f_i$$

with the weighting factors defined as quotients of mean sunspot area / mean sunspot number·

Waldmeier classification	A	B	C	D	E	F	G	H	J
weighting factor G_i	4	4	8	18	25	36	50	44	37

has been tested by a group of observers over several years. The results are compared with measurements of the projected sunspot area.

New classification schemes were developed for white-light faculae, light bridges, and prominences, as a base for long-term statistical investigations. Facular activity is measured by the number of faculae groups visible. A distinction is made between faculae accompanied by spots, and faculae without spots. First results based on observations from 1979 to 1986 show that the modulation in the total number of faculae groups is caused by faculae that surround spots (Fig. 3).

Measurements of sunspot positions allow activity in northern and southern latitudes to be distinguished. A significant surplus of spots at southern latitudes was found from 1983 to 1985 (Fig. 4). Synoptic charts of the photosphere show the distri-

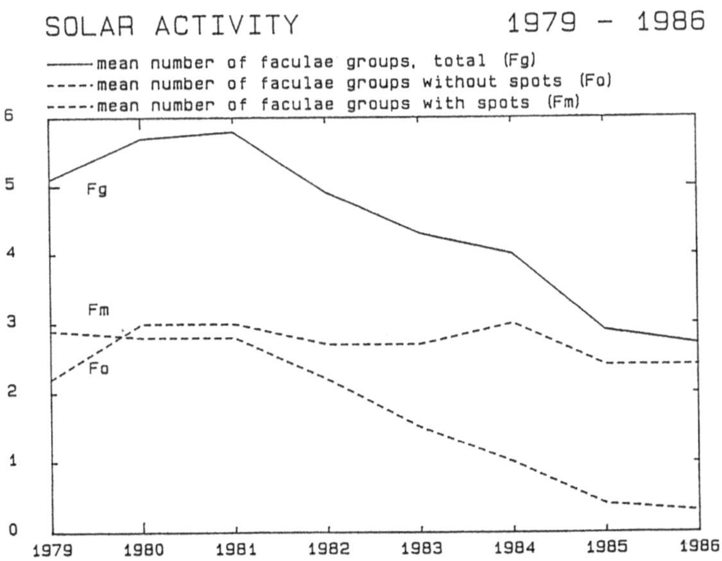

Fig. 3. Facular activity measured by the number of faculae groups visible (based on evaluations by V. Gericke)

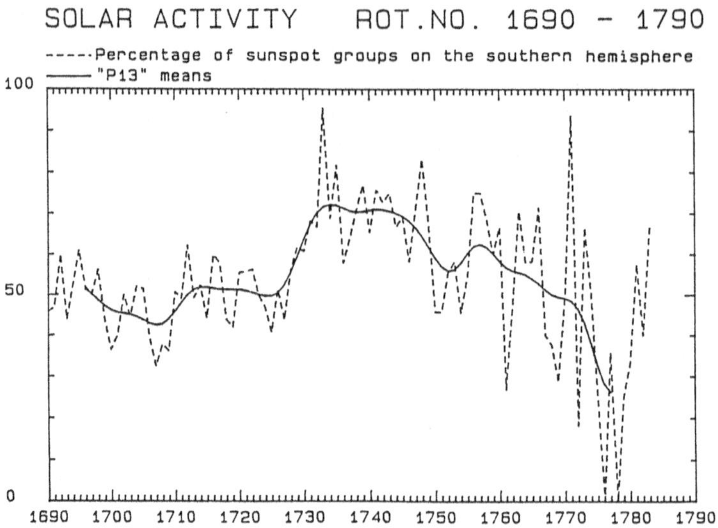

Fig. 4. Percentage of sunspot groups on the southern hemisphere (based on computations by E. Junker)

bution of spots in longitude and latitude. These charts are drawn for each Carrington rotation, and yearly as a summary diagram. The solar "butterfly diagram" published by professional authors until 1976 is continued by amateur work.

Photographic observations in white light, Hα and Ca II (K) are used, e.g. to simultaneously study the development of active regions. For example, the large sunspot group in June 1983 was analysed in detail, confirming a relation between proper motions within a group and flare activity.

Acknowledgements

I wish to thank Günter Appelt, Dr. Rainer Beck, Volker Gericke, Sieglinde Hammerschmidt, Cord-Hinrich Jahn and Elmar Junker for contributing diagrams and results of their work to this paper. I also acknowledge the kind assistance of Stefan Haacke who provided the French abstract for the conference programme.

The Moonwatch Program: A Model for Amateur Contributions to the ISY

James Cornell
Smithsonian Astrophysical Observatory, Cambridge, MA 02138, U.S.A.

Largely forgotten today is the role that amateur astronomy played in the early days of space exploration. For example, 1987 is not only the 30th anniversary of the first artificial satellite but also of history's most successful global organization of amateur astronomers: Moonwatch.

As a worldwide network of volunteer visual observers, Moonwatch was a vital adjunct to Western satellite tracking efforts. In nearly two decades of operation, more than 500 individuals made some 400,000 observations of approximately 6000 objects. Although now only a footnote to space history, Moonwatch suggests that some comparable organization of dedicated and talented amateurs might contribute to the International Space Year of 1992.

Moonwatch resulted from the International Geophysical Year of 1957-58. Fred L. Whipple, director of the Smithsonian Astrophysical Observatory, had earlier accepted United States responsibility for the optical tracking of any satellites launched during that 18-month period. As his primary instrument, he proposed a fast, large-aperture Schmidt camera called the Baker-Nunn. Eventually, 12 cameras would be located around the world.

Whipple also suggested establishing a volunteer network of observers to locate satellites visually and provide the basic data required for more precise optical tracking. A worldwide appeal for "Visual Observers of Satellites" was published in the July 1956 Sky and Telescope and planetarium director Armand Spitz personally contacted many astronomy groups. Response was immediate: some 1500 volunteers were recruited in the United States alone within the first few months.

In the spring of 1957, Smithsonian set up a professional steering committee chaired by George Van Biesbroeck, then of Yerkes Observatory. By that summer, after a series of test alerts proved a visual system could effectively track satellites, Moonwatch had 80 teams in the United States, 30 in Japan, 10 in both Argentina and Germany, and 33 more distributed over another 15 nations.

Typically, a Moonwatch team would set up a "fence" of observers using several wide-angle telescopes trained on the meridian at different altitudes. When an observer at one "picket" saw a satellite cross the meridian (defined by a cross bar on a vertical pole), he signaled the team's timekeeper, thus obtaining a fix on the object. The leader would then relay the information to the Smithsonian.

The basic Moonwatch telescope was of rugged, simple design with a 50mm objective, a 180mm focal length; and a 6.8-degree field of view. The optics were mounted in an 8.5-inch aluminum tube. The 6-power instrument could show objects down to 8th magnitude and the wide field increased the chances of finding a satellite even when its orbit was not well known.

The establishment of Moonwatch was done in an atmosphere of some urgency, since both the United States and the Soviet Union were on the verge of launching satellites. Indeed, when Sputnik I was launched on October 4, Moonwatch was one of the few Western organizations ready to provide tracking data. The first confirmed sightings came October 8 from teams in Sydney and Woomera, Australia; the first American observations came two nights after that.

During the initial weeks of the Space Age, essentially all observational data outside the Soviet Union came from Moonwatch. Eventually, Moonwatch would make 700 observations of Sputnik and its third-stage rocket. And, by the end of the first full year, volunteers had amassed nearly 7000 visual observations of the first seven satellites.

When IGY ended, both Moonwatch and Baker-Nunn networks continued under the Smithsonian with support from NASA. Naturally, the increasing precision of other ground-based systems significantly changed Moonwatch objectives. For example, Moonwatch was particularly well-suited for observing low-perigee objects and for tracking reentering satellites. The most spectacular success occurred in September 1962, when Moonwatchers observed the reentry of Sputnik 4 over the midwestern United States and subsequently recovered a 20-pound chunk of debris.

Although originally intended to operate for only 18 months, Moonwatch provided data for over 18 years, being disbanded finally in 1975. Most extraordinary was the dedication and persistence of Moonwatch members over those nearly two decades. They performed an arduous and time-consuming task with little reward beyond the knowledge that they were aiding international space goals. Moonwatch demonstrated well what amateurs could do when properly inspired and led.

On a practical level, Moonwatch was extremely cost-effective: producing nearly a half-million observations worth more than $14 million. Less tangible, but equally important, Moonwatch made significant contributions to the public understanding of science, especially in those hectic, and for some people, quite alarming, days of Sputnik. Also, many younger members, introduced to space science in this direct way, went on to pursue careers in astronomy, astrophysics, and related fields. And perhaps more than any other effort, Moonwatch served to underscore the truly international nature of space research.

These three contributions of Moonwatch – public information, science education, and worldwide understanding – are important considerations for any comparable amateur program mounted in the International Space Year of 1992.

Scientific Goals for a Collaboration Between Amateur and Professional Astronomers in the

Study of Variable Stars

Ennio Poretti
Osservatorio Astronomico di Brera, Via E. Bianchi, 46, Merate, Italy

Variable stars are a field where the contribution of amateur astronomers is of great importance. This contribution can be separated in to complementary parts:
1) collaboration with professional institutes that well-equipped amateurs can easily undertake in the field of photoelectric observation;
2) visual observation of variable stars, which in many cases can be a preliminary approach to photoelectric photometry.

Amateurs using their own photoelectric equipment can, of course, plan their work independently of astronomical institutions, but I think that their collaboration could produce interesting results. Here I would like to point to the advantages that a network of photoelectric observers can secure. I have appreciated these advantages when analysing photoelectric measurements carried out by both amateur and professional astronomers on 39 = AY Cet, the RS CVn star with largest known difference between orbital and rotational periods. With a period of 77 d, only the contribution of many well-equipped observers could secure the required light-curve coverage. Compare the wealth of information contained in our work (Poretti *et al.*, 1986) with the fragmentary light-curve of Fig. 1, where all the available measurements on V509 Cas are plotted together with those carried out at Merate Observatory in 1985 and 1986. The behaviour of this long-period massive supergiant is far from being understood and many questions remain unanswered. I would also like to draw attention

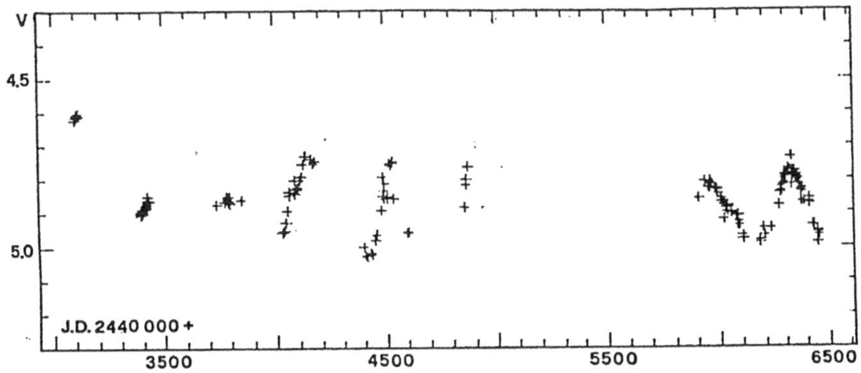

Fig. 1. Photoelectric light curve of V509 Cas from 1977 to 1986

to yet another possibility offered to a group of observers with a spread in longitude: the continuous monitoring of a short-period variable star. The advantages are essentially two: 1) the reduction of aliases in frequency analysis; 2) the opportunity of analysing cycle-to-cycle variations. Multi-site observations allowed us to solve the complex light-curve of the δ Sct variable ϑ^2 Tau ($\Delta V = 0.04$ mag; Breger *et al.*, 1987). A data analysis does not show any discernable alias structure, thus allowing an easy period identification: this is not the rule for δ-Sct-star frequency analysis. The double-mode pulsator V1719 Cyg is another example: in this case it has been sufficient to link two sets of observations made in Italy and in the U.S.A. to obtain a clear indication of the second period (Antonello *et al.*, 1987). Extensive photometry of BW Vul (Sterken *et al.*, 1987) proved that the standstill on the ascending branch of the β Cep variable is stable, thus sweeping away conjectures on its variability (in phase and length) based on insufficient data.

It is the general opinion that visual investigation of faint E/RR variables can yield satisfactory preliminary elements for a more accurate study: in a way this can be seen as a preparatory approach to photoelectric photometry. Indeed, the AAVSO is known and acknowledged all over the world for its survey of long-period and cataclysmic variables. However, it is of importance to note that many associations now use automated routines to acquire and process data. For this reason, a general improvement in data reduction is required to provide reliable light-curves for red, small-amplitude (< 0.5 mag) variable stars. In fact, I do not think that a photoelectric survey, even if preferable, could be undertaken of all known bright variables. Ralincourt *et al.* (1987) emphasized the existence of effects that may alter visual light-curves and they developed a method that gives more satisfactory results by making an appropriate correction. The analysis of estimates secured by GEOS on five SR variables shows that an average observer can make estimates with a standard deviation of 0.08 mag, but is affected by a systematic error of the same order.

References

Antonello, E., Mantegazza, L., Poretti, E., 1987: in *Lecture Notes in Physics*, No. 274, p.191
Breger, M., Antonello, E., Mantegazza, L., *et al.*, 1987: *ibedem*, p.122
Poretti, E., Mantegazza, L., Hall, D.S., *et al.*, 1986: *Astron. Astrophys.*, **157**, 1
Ralincourt, Ph., Poretti, E., Boistel, G., 1987: *Astrophys. Space Sci.*, **134**, 135
Sterken, C., Young, A., Furenlid, I., 1987: *Astron. Astrophys.*, **177**, 150

The Contribution of Amateur Astronomers to the Study of Variable Stars

Dominique Proust[1] and *Emile Schweitzer*[2]
[1]Département d'Astrophysique Extragalactique et de Cosmologie, Observatoire de Meudon, F-91295 Meudon Principal Cedex, France
[2]Association Française des Observateurs d'Etoiles Variables, F-Strasbourg, France

The observation of variable stars is one of the most important and the most fruitful areas of stellar astronomy. The contribution of large numbers of visual observations is a determining factor in drawing up light-curves, the latter being the key to the interpretation of the process of variability within these stars.

It is not easy for a professional astronomer to obtain access to modern telescopes, especially to those of large aperture, and it may even be difficult, given the large number of projects put forward. In any case, these large instruments are often unsuitable for the observation of variable stars. Amateurs, on the other hand, have instruments that have a lower degree of precision, but their greater number and the good organization that exists for the reduction of data obtained, represent trump cards in preparing light-curves.

The amateur thus has a wide gap that can be exploited by using a modestly-sized telescope in conjunction with a physiological, rather than physical, detector, the eye, which is capable of making measurements, which are of sufficient accuracy for most types of variation to be scientifically useful.

The development of astronomy from space has also opened up new observational possibilities (with the IUE and IRAS satellites, for example); variable stars radiate in most of the principal regions of the spectrum, so it is possible to make coordinated observations, the visible region being reserved for the amateurs.

Thanks to observations made by the members of the Association Française des Observateurs d'Etoiles Variables (AFOEV) and linked to well-established scientific programmes (Schweitzer and Proust, 1987), numerous results have been obtained in the last few years, over a whole range of types of variation.

Mira-Ceti-type Stars

The intrinsic properties of these stars (long periods and large-amplitude variations) make them ideal candidates for amateur observation. The light-curves established (Fig. 1) enable the following studies to be carried out:

- Correlations between the magnitude and the variation of other spectral or photometric characteristics (OH, H_2O, SiO masers, cf Fillit et al., (1977); study of the radial velocities of the absorption lines).

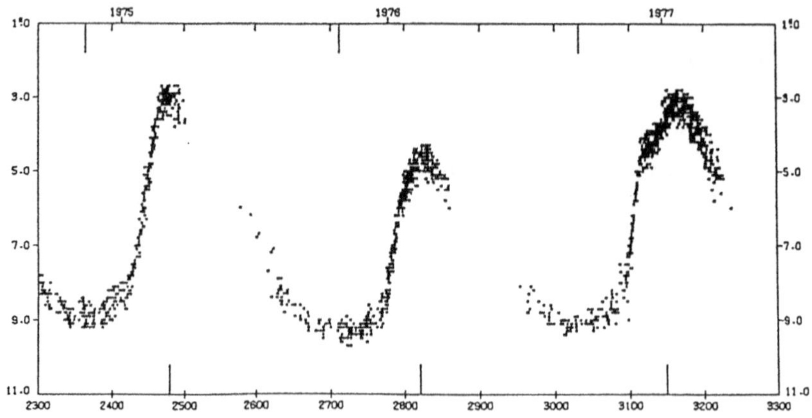

- Use of the parameters of the light-curves, averaged over a large number of cycles (statistical links between these parameters and the final evolution to the white-dwarf, or planetary-nebula stages).
- The physics of Mira stars by studying the cycle-to-cycle variations (stellar structure, dynamical evolution, mass-loss: cf Wood, 1982).
- Fourier analysis of the light-curves and establishing ephemerides for solar (Speckle) or space missions (Hipparcos):

$$m(t) = m_0 + \sum_{i=1}^{3} A_i \cos(2\pi\nu_i(t - t_0 + \phi_i)$$

where $i = 1$ corresponds to the principal mode ($\nu_1 = 1/p$), p being the star's primary period;
$i = 2$ is this mode's harmonic ($\nu_2 = 2/p$);
$i = 3$ corresponds to the long-term variation that is shown by almost all Miras when their light-curves are analyzed by Fourier methods;
m_0, A_1 and ν_1 are taken from the values given in the GCVS (Kukarkin et al.);
ϕ_1 is taken to be equal to π and t_0 to the date of the last maximum observed (Mennesier, 1987).

Other Variables

The observation of variable stars is not, of course, limited to Miras; the study of stars at the end of their evolution that are found in the centre of planetary nebulae (Acker and Jasniewicz, 1985), is a fundamental tool in understanding the processes that lead to the formation of planetary nebulae. Where *novae* and *supernovae* are concerned, amateur astronomers have a vital role, because their light-curves are precious tools in understanding the physics of these objects, as well as the major role of supernovae in the chemical evolution of galaxies, and their enrichment in metals and radioactive elements. The observation of Z-Cam-type stars enables us to refine our models of accretion disks that are required to explain certain features of the light-curves in the standstill phase (short duration and fluctuations of minima). The phenomena that occur in symbiotic stars are still too poorly understood for us to explain fully the transfer of material in close systems.

All these examples clearly illustrate the vital role played by variable-star observers. Professionals involved in stellar astronomy more and more frequently request observations. The increasing number of simultaneous observations brings larger and larger forces into play, far transcending national boundaries, in particular by mobilizing amateur and professional observers of variables. Despite this cooperation, a large number of questions about variability remain unanswered or have only partial answers. Amateur astronomers can help to solve these problems.

References

Acker, A., Jasniewicz, G., 1985: Astron. Astrophys., **143**, L1

Fillit, R., Proust, D., Lepine, J.R.D., 1977: Astron. Astrophys., **58**, 281

Kukarkin, B.V.: *General Catalogue of Variable Stars*

Mennessier, M.D.: 1987, *INCA report*

Schweitzer, E., Proust, D., 1987: L'Astronomie, **101**, 303

Wood, P.R., 1982: *Pulsation in classical and cataclysmic variable stars*, eds. Cox, J.P. and Hansen, C.J. (Joint Institute for Laboratory Astrophysics, Boulder, Colorado)

Professional/Amateur Cooperation in the Study of Variable Stars as Illustrated by the Fourth Edition of the GCVS

M.S. Frolov
Astronomical Council, USSR Academy of Sciences,
48 Pyatnitskaya ul., 109017, Moscow, U.S.S.R.

Let us divide variable stars into two main groups: the first "classical" group, includes objects known for a long time, such as Cepheids, RR-Lyrae stars, Miras, cataclysmic variables, eclipsing binaries, etc. The second group includes micropulsating variables of δ Scuti and β Cephei types, magnetic variables, rotating variables of BY Draconis type, etc.

Historically, the contribution of amateurs in investigating the first group was very significant, and it continues to increase. On the other hand, involvement in studying the second group of stars was practically equal to zero some years ago, but today one can see the beginnings of an expansion of amateur work on this second group of variables – among brighter objects, of course. One reason is the beginning of cooperation between amateurs and professional astronomers having powerful instruments.

Another good example of the fruitful collaboration between amateur and professional astronomers over many years is in the preparation of the *General Catalogue of Variable Stars* (the GCVS). This *Catalogue* is well-known around the world, not only to variable-star investigators, but now also to a considerable number of modern astrophysicists.

Since the publication of the first edition of the *Catalogue* in 1948, two other editions have been prepared and fourteen *Supplements*, three editions of the *Catalogue of Suspected Variables*, twenty-five *Name-Lists of New Variable Stars* and three volumes of the fourth edition of the GCVS. The fourth edition, including the NSV-catalogue, will contain information of fifty thousand variable stars in our Galaxy and in external galaxies.

The team of authors of the GCVS makes wide use of the data on variable stars obtained by amateurs. Among the principal bibliographic references in the GCVS we can find such amateur publications as the *Journal* of the AAVSO, *AAVSO Abstracts*, *Bulletin* of the AFOEV, publications of GEOS, RASNZ, BAA VSS, publications of Italian amateur astronomers, and BBSAG (Switzerland). We continue to urgently need maxima of Miras, traditionally published by the AAVSO, minima of eclipsing binaries (BBSAG), maxima of RR-Lyrae stars from different amateur sources, results of regular studies of cataclysmic variables from AAVSO and RASNZ bulletins. For example, the contribution of amateurs in the field of Miras amounts to about 90 %.

Today there are few amateur observations that we can use for variables in the second group, because microvariables can be successfully investigated only with high-precision photoelectric data. Only a wide distribution of photoelectric photometers and CCD-photometers among astronomers can seriously change this situation. In this respect we see very good prospects from the cooperation between professionals and amateurs. A good example of such cooperation is the IAPPP.

The team of authors of the GCVS is very indebted to all the amateur astronomers for the useful data that help us to prepare our *Catalogue*.

Amateurs and the Search for Supernovae

Stephen H. Lucas
14400 S. Kolin Ave., Midlothian, IL 60445, U.S.A.

The role of amateur astronomers in the search of extragalactic supernovae is emerging. Visual discoveries by John Bennett in 1968 and Gus Johnson in 1979 were isolated and surprising encounters, but are indicative of the possibilities. Between 1980 and the present, these possibilities have been well-tested by the Rev. Robert O. Evans, who has discovered 15 SN's and shares credit for a 16th. Rev. Evans' record was achieved through an organized approach to observation and assisted by the development of a unique observing regimen. Detection of SN is a formidable challenge, but the obvious conclusion is that amateurs may expect to attain more than marginal success. We approach this subject from the standpoint of visual observation, but are obliged to acknowledge the contemporaneous and successful application of photography as a search tool.

Steps that must be taken to improve the likelihood of visual discoveries depend on *persistance, method, charts,* and *equipment*, and it should be no surprise that *motivation* is a major guarantor of this undertaking. Amateur discoveries have contributed to the resurgence of general interest, especially within the past few years. They have contributed to the current flowering in the theoretical analysis of Type I SN, led by Profs. J. Craig Wheeler, David Branch, and others. Especially, two of these discoveries have led to the realization that there are two completely different classes of SN with a Type I light-curve. The new class is now called Type Ib, and the two SN 1983N and 1984L have been prototype examples of the class. The two brightest SN (1983N and 1986G) have also been used as probes of intergalactic space, and have resulted in the discovery of several "clouds" of gas, between the Milky Way group of galaxies, which had not been known before. There will be effects from this discovery in the study of quasars and other very distant objects.

With the proper organization, amateurs will be able to make a major contribution in deciding the rate of appearance of SN in the various types of galaxies. By collating and studying statistics of negative an positive search observations on many different galaxies, and over an extended period of time, we can know more clearly the average SN rate. An article recently submitted to *Astrophysical Journal* made a study of 50,000 visual observations of 1081 galaxies over a 5-year period. A sub-sample of 709 Shapley-Ames galaxies was used to help create homogeneity. Of the 29 SN discovered world-wide in these galaxies during the 5 years, eleven were discovered by the visual observer, and 4 others were also seen, resulting in a total of 15 events falling within the scope of the search. Estimates of the SN rate were made on this basis. (Van den Bergh, *et al.* 1987)

The detection of these enormous stellar events represent one of the final bastions of scientific discovery left open to the amateur. Careful scrutiny is the price that has to be paid for such a project. Our group SUNSEARCH is not like most groups where the few do the work for the many. We are the few, and have elected, by intent, to participate in a tedious program of amateur astronomy. It certainly isn't the easiest specialization, but it can be rewarding; if not, then just making us better observers and perhaps making a new friend along the way. After all, isn't that what it's all about anyway?

The U.K. Nova/Supernova Search Programme

Guy M. Hurst Coordinator
16, Westminster Close, Kempshott Rise, Basingstoke, Hants, RG22 4PP, U.K.

A summary of the results of the UK Nova/Supernova Patrol is presented together with a brief outline of projects undertaken by our members.

The UK/Nova/Supernova Patrol was formed in 1976 June and the project is run on behalf of the British Astronomical Association, The Astronomer magazine and various overseas groups with the aim of searching for novae, supernovae and to monitor various 'recurrent' objects.

The early successes in 1975/1976 were principally the finding of various pre–discovery images of novae and other eruptive objects on routine patrol films, e.g. : V400 Persei, V373 Scuti, HM Sge. This provided valuable information on the light–curves of such objects and helped to determine the date and magnitude of maximum.

The first major success for the patrol came in 1977 January when John Hosty discovered HS Sagittae (= Nova Sge 1977). The equipment was modest 10 x 50 monocular and from a light–polluted area. This emphasized that expensive equipment was not needed to be successful, just dedication from the observer.

D. Rossiter was unlucky in 1978 when he discovered V1668 Cygni, but after an earlier discovery in USA. M. Swan found pre-discovery images of PU Vul on patrol films in 1979, and John Hosty rediscovery CSV 101897 on films in 1980 October.

The most mysterious patrol discovery occurred in 1981 January when D. Branchett found a nova-like object in Scutum. Although not subsequently confirmed, research showed that the star had been noted in earlier records as an 'atlas omission'. Further deep photographs failed to show a new object to a limiting magnitude of 18.

The star VY Aquarii, then classified as a recurrent nova, featured in patrol news of 1982–83. In 1982, R. McNaught announced he had found the star of the Papadopoulos Atlas and followed this by observing it at maximum in 1983 November, the first time it had been seen visually. This, and subsequent research by patrol members, led to the star being re–classified as a dwarf nova.

Annick Merlin in France recorded PW Vul (= Nova Vulpeculae 1984) on photographs prior to its discovery announcement 1984 August. Then, in 1985 April, H. Mikuz, operating patrol cameras in Yugoslavia, recorded an outburst of RS Ophiuchi.

The trend towards monitoring 'recurrent objects' continued and success was achieved with the first-ever visual observation of DO Draconis by S. Lubbock in 1985 October.

A further major success for the patrol came in 1986 with the discovery of Nova Centauri 1986 by R. McNaught, one of our patrollers in Australia.

The same observer provided a pre-discovery series of pictures of the Supernova 1987A in the Large Magellanic Cloud. These results, as with the numerous previous photographs of newly discovered objects, provided invaluable information on the object's rise from minimum.

He followed this with another nova discovery in Sagittarius in 1987 May, the third nova discovery for the patrol.

The above notes provide just a summary of the main achievements but are by no means complete. A detailed report on the results of the project for 1976–1987, together with a full list of bibliography references, will appear in a future issue of the Journal of the British Astronomical Association.

If any observers wish to participate in the search for novae and supernovae, or astronomers require details of our results, please write to the author for further information.

GEOS – Groupe Europeen d'Observations Stellaires

Michel Dumont
3 promenade Venezia, F-78000 Versailles, France

GEOS was formed in 1973 from several European groups (French, Italian, Spanish and Belgian). All the members of GEOS have the same aims: to observe variable stars and to analyse their observations themselves.

Other objects are also observed by GEOS: artifical satellites and minor planets, in particular minor-planet occultations. Observations are treated statistically and the procedures allow accidental, and – if there are several observers of the same star – systematic, errors to be corrected, thus improving the accuracy of the results and the light-curves.

Initially, therefore, GEOS encouraged members to accumulate a large number of observations with a view to their later reduction. Up to 1980, the methods were visual, using the Argelander method. GEOS receives about 100 000 observations every year and their analysis was quickly decentralised. After 14 years, more than 500 papers have been published.

Although GEOS members observe a wide range of stars, certain categories have emerged as being particularly important:

1) Pulsating variables, where the aim is to obtain the mean light-curve, to confirm the ephemeris or to determine the O−C, an indication of a slight change in period. The short-period, RR Lyrae stars are well observed, especially the Blazhko effect, which indicates a double period.

2) Eclipsing variables are observed in the same way. In addition one tries to directly determine the times of minima, and to calculate the orbital elements from the light-curve.

3) Favourite among GEOS members, however, are little-known stars, where there is a possibility of quickly discovering something new. These stars are essentially of two types, low-amplitude variables and variables that are always faint, with maxima below magnitude 10. These stars need a telescope of more than 20 cm to be properly studied.

When an amplitude is greater than 2.5 magnitudes, observational errors can be considered as negligible in comparison with the observed variation. Between 1 and 2 magnitudes, particular care must be taken in making estimates and the detection of errors becomes useful. Below one magnitude, not only should the estimates be made with the utmost care, but statistical treatment of the observations becomes essential. Such reduction allows spurious points to be eliminated and a smooth light-curve to be drawn. When several observers have carefully followed the same star, systematic errors can be eliminated. Correction of systematic errors must always be made before calculation of the final mean of observations by several observers.

Statistical treatment has allowed the range of visual observations to be extended, and for their potential to be established. Variation of one magnitude is still detectable, and if the variable is periodic, visual estimates allow the variation to be detected if the amplitude is more than 0.2 magnitude. If the variations are not periodic it is difficult to draw any conclusion about them below 0.4 magnitude.

The procedure for studying new objects is that they are chosen, often from the NSV, and usually for showing rapid variations. A visual observational campaign follows, in particular monitoring the star through the night and over several nights. Each observer obtains a set of points to which he tries to fit a periodic curve. The best algorithm is to produce a periodogram, which will show the probable period. From the light-curve and period an attempt is made to determine the type of variation and a preliminary ephemeris. Any new results obtained by visual methods should be confirmed. GEOS therefore organises photoelectric observing runs at an observatory (Pic du Midi, Jungfraujoch, etc.). The results are published either in the GEOS *Circular* or in the *IBVS*.

M. Roland Boninsegna will describe GEOS results for a specific star.

GEOS Results for PX Cephei

Roland Boninsegna
rue de Mariembourg 33, B-6381 Dourbes, Belgium

This star's variation was discovered photographically by Romano in 1957. It was included in the GCVS in 1978 as PX Cep, and I started to observe it in 1983.

The star is usually at maximum, but on several occasions I was able to observe either a rise or the start of a fall. Finally on 1984 October 21, the first minimum was observed, and then a second one ten days later.

The star's constant behavior, outside the minima, indicated that is was an EA type of eclipsing binary. The amplitude of the eclipses is about 2.6 mag. Secondary minimum therefore has an amplitude of 0.11 mag., which is undetectable visually.

The first minima observed, 9.376 days apart, were a multiple of the true period P. As the period is obviously longer than one observing night, there are few possible periods: 9.376 days; 4.688; 3.125; 2.344, etc. An ephemeris was calculated for all the possible periods and attempts were made to observe the predicted minima. Finally only the 3.125-day period remained in the running. Using it, an initial ephemeris was calculated, and this was confirmed by the use of the 76-cm photometric telescope at the Jungfraujoch on the trips organised by the Palais de la Découverte in 1985 and 1986.

Photoelectric observations show that PX Cep varies between 12.25 and 14.65. The B−V index varies from 0.32 to 0.95: the pair probably consists of a main-sequence A star and a K giant. The secondary minimum was not detected.

My visual estimates over several years have allowed me to determine a farily accurate period: $P = 3.126993$ days ± 0.000006 day.

The French Network for Observing Comets and Small Solar-System Bodies

Robert Chemin
Téléscope de Schmidt, CERGA, Caussols,
F-06460 St. Vallier de Thieu, France

For the apparition of P/Halley the CNRS set up RCP 639 (Cooperative Research Programme), one part of which was to coordinate French amateur work. This carried out a training programme, at first by publishing (with help from ESO) an observational manual (eds: J.L. Heudier, J.C. Merlin and M. Festou), and then by a series of courses covering astrometry, photography, visual magnitude estimation, and drawing of jets. These courses were provided by J.L. Heudier, J.C. Merlin and T. Laverge, and were taken by about 100 observers. The author was asked to undertake the organization of the section so that it might become a permanent network of observers of comets and minor bodies.

At present the network comprises more than 220 participants, clubs and individuals, grouped into 19 regions. A meeting bringing together about 30 of the best amateur observers was held at Calern on June 6–8, and the observations obtained for the apparition of Comet Halley have been collated with a view to their publication by the CNRS. Those present at the meeting, and many others, have asked that the network be kept in being. With some restructuring, the CNRS will be asked to form it into a strictly amateur Cooperative Research Programme.

The network has important assets (and can obtain others). These include: a) about one hundred trained observers, with an observing manual and positional reduction programmes; b) an observatory at Plan de la Tour (in the Var), with a 300×500-mm telescope built by REOSC, which is capable of reaching mag. 18 with a 9° field, and a photographic laboratory. Information about use of the telescope can be obtained from D. Albanese, 103 rue Jean Aicard, Fréjus (Var).

If the network is continued, it will be able to use IT methods currently under development to enable anyone to access programs for calculating ephemerides, and identification of minor bodies in the Solar System, that are used for the Schmidt at Calern. This will be possible only with the help of CNRS.

At the meeting it was decided to continue training courses, to prepare new project lists, more suitable for each type of activity, and finally, to appeal to the major amateur organisations to help in the network's activity and thus enable it to form a part of international observational work.

Amateurs' Contribution to the PHEMU85 Campaign Observing Mutual Phenomena of Jupiter's Satellites

J.E. Arlot
Bureau des Longitudes, 77 avenue Denfert-Rochereau,
F-75014 Paris, France

Mutual events (occultations or eclipses) of Jupiter's satellites occur every 6 years. The 1985 series was particularly favourable as it occurred when Jupiter was at opposition. An observing campaign (PHEMU85) was therefore organized by the Bureau des Longitudes, and sponsored by the CNRS under RCP (Coordinated Research Programme) 754.

The interest in observing the events lies in the precise positional information that may be obtained. Conventional (photographic) astrometry gives results to within 400 km, but mutual events give an accuracy of at least 100 km. Observation of mutual events is easy in that the Galilean satellites are bright, so large telescopes do not have to be used, and photoelectric observations in particular can be obtained from urban sites or under poor meteorological conditions. The professional workers generally employed photoelectric photometers or vidicon-type detectors, while amateurs used visual, photographic, photometric and video techniques. Two groups of amateurs were particularly involved, GEOS (European) and GEA (Spain), but individual amateurs also contributed.

The observations essentially consist of recording the magnitudes over the duration of the events (about 10 minutes), and the time (UTC). Although photoelectric observations gave the highest precision, some visual light-curves have sufficient accuracy for them to be used in future theoretical studies. Occultations, where the amplitude is small, were difficult to observe visually, but were detected.

Photographic observations used standard 35-mm film (Ilford HPS), a 30-cm telescope and 4-second exposures, and obtained scientifically interesting results.

We hope to mount a similar campaign in 1991 and invite amateurs interested in participating to contact us as soon as possible.

Wanted: An International Jovian-Satellite-Phenomenon Centre

Niels P. Wieth-Knudsen
Svend Trostvej 12 IV, DK-1912 Fredriksberg C, Denmark

Abstract

Amateur observations made since 1977 show that although the ephemerides for the satellites of Jupiter published by the Jet Propulsion Laboratory are more accurate than those in the *Astronomical Almanac*, there are still small deviations, which may increase with time. Observation of eclipses, transits and occultations, as well as of the mutual phenomena that occur at 6-yearly intervals can provide information for correction of the ephemerides. There appear to be three groups coordinating observations: 1) in Germany, started by P. Ahnert in the early 1960s and now under H.-J. Blasberg; 2) in the U.S.A., started by J. Ashbrook in 1976, and continued by J. Westfall of the ALPO; 3) in Australia and New Zealand, led by B. Loader. These groups appear to work independently, and it would greatly help research on this subject if there were an International Centre for collecting observations of the Jovian satellites – perhaps under the supervision of the IAU – whereby all observations made anywhere in the world would be available to anyone investigating the subject. Paris, where the predictions are calculated, might be a very suitable site for such a centre.

The European Section of IOTA: Scientific Programme Coordination and Results

W. Breisker and *H.J. Bode*
IOTA/ES, Bartold-Knaust-Str. 8, D-3000 Hannover 91, Fed. Rep. of Germany

Abstract

Lunar occultations and occultations of stars by planets and asteroids are important tools in determining astrometric information as well as details about the occulting bodies. Successful work requires a relatively large number of small observatories, well-distributed over a large area of the Earth. This is particularly important in occultations by asteroids because the accuracy of the predictions is generally poor.

Cooperation is therefore essential. IOTA/ES (European Section of the International Occultation Timing Association) provides a means of cooperative work, and holds meetings in different European countries about once a year. Eastern European countries are included.

Some results of occultations were presented and future events discussed. A relatively cheap photoelectric system was described, very suitable for amateurs and public observatories.

[Text not received. – Eds.]

The Role of the Amateur in Modern Astronomy

H. Derek Howse[1], *Patrick Moore*[2]
[1]12 Barnfield Road, Riverhead, Sevenoaks, Kent TN13 2AY, U.K.
[2]Farthings, West Street, Selsey, Chichester, W. Sussex, U.K.

Abstract

This contribution dealt with the need for international coordination between amateur astronomers and the failure of the IUAA (International Union of Amateur Astronomers) to meet this need. [The substance of this contribution was given in *Sky & Telescope*, **73**, 482 (1987 November).The text was not received. – Eds.]

Part IV

Popularization

Meeting the Public: Popular Observatories

Josip Kleczek
Astronomical Institute, Czechoslovak Academy of Sciences,
25165 Ondřejov, Czechoslovakia

"L'homme est infiniment petit par son corps,
mais il est infiniment grand par son esprit".
Blaise Pascal

1. The flow of information and popular observatories

Popular observatories (PO) are an important link between astronomers and the public. Fig.1 represents the flow of information about the Cosmos and the role of POs.

The flow of information about celestial bodies starts with photon emission. The emitted photons carry the information about the moment when they are born. They transfer the encoded

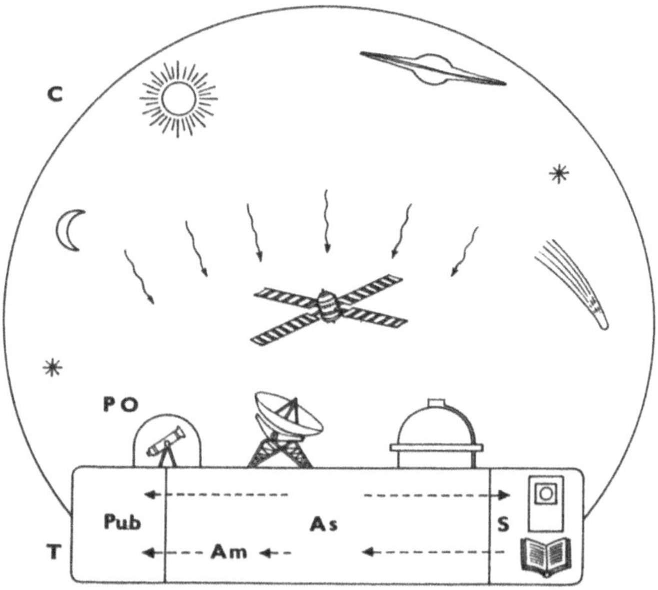

Fig. 1. Photons (corrugated arrows) transfer information from the Cosmos (C) to terrestrials (T). The information is received and decoded by observational instruments. By astronomers' minds it is interpreted and integrated into a system of knowledge. The flow of astronomical knowledge in the community of terrestrials is shown by dotted arrows. It is in the minds of astronomers (As), or stored in books, journals and on magnetic tapes (S). It is communicated to amateur astronomers (Am) and to the public (Pub). Popular observatories (PO) are an important interface between astronomers and the public.

information through cosmic space in all directions, among others towards our planet. By ground-based or space-based instruments astronomers decode the information in the photons. The size and shape of celestial bodies are determined with great precision (by photography and interferometry). Direction from which the photons arrive determines the position of their source (astrometry). The amount of photons (photometry), their frequency (spectrophotometry) and the direction of their oscillation (polarimetry) are other information decoded from the incident photons. The observational instruments are the interface between the Cosmos and astronomers' minds.

Observational data represent the fundamental information for our knowledge of the Cosmos. For the astonishing achievements of astronomy, however, we owe much more to the logical reasoning of the scientists who interpret the data. Our knowledge of the Cosmos is based on two pillars : the observational data decoded from photons and the ordered thought of the human mind. The flow of astronomical knowledge within our terrestrial civilization is marked in Fig.1 by dotted arrows.

The knowledge may be stored (S) to be retrieved later. It is communicated to amateur astronomers or the other way round. Both the astronomers and amateurs (As, Am) communicate the knowledge to the public (Pub). Who is the public ? Undoubtedly not every terrestrial belongs to the public in our sense. The greek word for man is *anthropos* which means "the one who turns one's head upwards". There are some terrestrials who turn their head upwards only to drink. They are rather rare and do not belong to the public in this context.

There are various channels of communicating knowledge to the public : radio, TV, newspapers, popular journals and books, lectures, planetaria and popular observatories. In particular, popular observatories represent a very active and the most efficient interface between astronomers and the public. In a PO one <u>listens</u> to lectures and discusses various problems; one <u>sees</u> movies, models of celestial bodies and of their systems, or one can see them with a telescope; and most important of all, one can <u>do</u> astronomy under the supervision of experienced amateurs and astronomers. There is a Chinese proverb : "I hear–I forget, I see–I remember, I do–I understand". Of all three activities at POs, the last one is the most efficient. Regular visitors of POs often "get their hands dirty on astronomy" to understand the structure of the Cosmos. They become amateurs and sometimes professional astronomers. After all, every amateur or astronomer has been recruited from the public. New-born amateurs or new-born astronomers do not exist.

2. Popular observatories in Czechoslovakia

Czechoslovakia is a small country. But we have more than 70 POs – autonomous institutions in their own buildings, with domes, telescopes, a conference hall and a library. Their staff is 2 – 5 (in small POs) up to 15 – 22 (in large POs). The fundation of our Astronomical Society, popularization of astronomy and construction of POs were stimulated by the Société Astronomique de France and Camille Flammarion, whose books were translated into our language. We owe much to them.

Activities in the POs consist in : 1) popularization of astronomy; 2) practical training in observations, data reduction and interpretation; 3) systematic observations in cooperation with academic and university observatories; 4) regular courses of lectures at secondary-school level; 5) two-year courses (3–4 days per month) at university level in POs at Hurbanovo and Valašské Meziříčí; 6) special activities such as the grinding of mirrors, the construction of telecopes, meteorological and seismological observations, tours with the 'astrobus' (it is a vehicle with telescopes and binoculars to observe the Sun and stars, movie and slide projectors; the driver is an astronomer), 'Bicycle tours' of astronomers and amateurs on bicycles (one week in the summer vacation going from one PO to another, with lectures and discussions), etc... Whenever I go to a PO, I am overwhelmed by deep admiration of the human will and impressed by noble perseverance. From what does their enthusiasm for astronomy stem ? Why do they spend so much effort to know more about the Cosmos and to share their knowledge with others ? What is the final value looked for ?

It is joy that comes from grasping the structure and beauty of the Cosmos, I presume. Thinking and understanding is the noblest human activity which brings joy and happiness : 'Felix qui potuit rerum cognoscere causas' (Vergilius)*. When composing the few bars, full of tension, in the Ninth Symphony, Beethoven had in mind the creation of the Cosmos; after the bars, the chorus sings the 'Ode in Praise of Joy'.

The joy of understanding and of sharing the knowledge with others is the final reward for all astronomers and amateurs who meet the public in popular observatories and elsewhere. Or do you have a better explanation for their enthusiasm and unselfish effort ?

---

* 'Happy is he who is able to know the causes of things '. – Eds

Comet Halley: A Media Event

Jean-Louis Heudier
Téléscope de Schmidt, CERGA, Caussols, F-06460 St. Vallier de Thiey, France

Comet P/Halley's life as a "media event" started long ago, because the first use of communication (in the modern sense of the term) goes back to Rabelais, who published in *Patagrueline Prognostication pour l'an 1532*:

"And whyth the comete sene laste yeare and the retrograte of Saturne a greate skoundrell shalle die, all weazynge and courd in scabbes, in the hospitall, and at his death there will be horrible warre between cattes and rattes, dogges and hares, faulcons and duckes, monkes and egges."

Since then ways of spreading ideas have progressed enormously, but comets have remained "fashionable" and each of the last six apparitions of P/Halley has been the occasion of outbursts that are far removed from the scientific view.

What appears important is that the cometary phenomenon is so easily used by the media. The spectacular and unexpected nature of apparitions are partly the cause, but the cultural impact of comets that have punctuated history also seems necessary to explain the persistence of fanciful ideas. Comets, despite scientific progress, have retained their mysterious, sacred and supernatural nature. You only have to glance through contemporary press cuttings of the most spectacular apparitions to be convinced of that. Nothing has changed from ancient fears. Contrary to what Alain[1] hoped at the beginning of the century, the "marvel" has not yet become a "tangible thing". People nowadays still like to frighten themselves, just as they did thousands of years ago. Did not Gassendi, three centuries ago, write:

"Yes, comets are frightening, but only because of our foolishness. We make our own objects of fear, and, not content with real ills, we acquire imaginary ones."

Exploitation of comets has progressively changed and we tend to laugh at the things that people frightened themselves with, but often the laugh is rather forced. Comets remain interesting and worrying because they have retained some of that mystery that mankind revels in. Over the course of centuries comets have caused fear for completely opposing reasons: at first because no-one knew what they were, and then because their movement and their composition was understood. The fear remains, only its cause has altered. And fear has always been an instrument of power. Fear is fed by ignorance: maintaining ignorance creates dependence and hence power.

[1] Emile Chartier, known as Alain, (1868–1951), French philosopher – Eds.

Helped by progress in communications, the phenomenon grew in a spectacular fashion in 1910, and with this first apparition in the 20th century, we can truly speak of "media manipulation". The comet was used in the very typical "small ads" of the time. Products connected with the event began to be sold: books, songs, postcards, gas masks, and "anti-comet" pills. Illustrations even appeared predicting trips to the comet. The phenomenon is important and Alain makes one of his most significant remarks about it.

"I can imagine what a historian one hundred years from now will write about the Great Terror of 1910. I can see him searching in our newspapers for everything that bears upon the comet. The very collection of documents about the same event will be in itself an error, because he would see remarks about the comet separated from everything else and forming their own little world. Naturally he will not pay attention to the ordinary, rational, way of life that does nevertheless go on around, and amongst, us, as everyone can see. Chroniclers only note what is strange and outside the ordinary. A six-page paper that wanted to leave future historians with an exact picture of human life on one day would have to fill its columns with: "People worked, drank, ate, slept; everyone thought about their own concerns and loves; there were births, deaths, illnesses, and follies, just like every year at this time. All's well." In such a description, and giving everything the space that is its due, doubtless the newspaper would not have a single line about all the crimes, extravagances and panics. Because humanity is wonderfully wise, and has perhaps always been; but the historian, perforce, sees it as mad, stupid, or bloody."

The press, by its very nature, greatly amplifies the real impact of comets. This impact is changed and exaggerated yet again by historians. We may ask how our own attitude during this last apparition will be regarded by our descendants in 2061. What is certain is that progress in communications since 1910 has widened the gap between the public's real attitude and the image given by the press. With all this "progress" the 1986 apparition just had to be the subject of media excesses: and it was.

Each of the latest passages of P/Halley has been the occasion of scientific "firsts", and the 1986 one was no exception: first electronic recovery, first space rendezvous, etc. The press therefore had a whole range of subjects with which to fill its columns, and it did not let the opportunity pass. Every paper in every country used the comet at least once to waste ink and paper. Thanks to all this clamour, new sales opportunities were created: special flights allowed people to (faintly) see the comet; cruises took astronomers and the curious to discover the sky; objects of all sorts were sold: the media event became a commercial event. Never have so many small telescopes been sold, and doubtless never have so many people looked at the sky: quite apart from proper scientific work, we can be quite certain that in the last few years there have been more people curious about the sky, and more telescopes pointed at the comet than over the whole course of astronomical observation.

Should we despair over such banal exploitation of a scientific event? Personally, I think not. I believe instead that de-facto situation ought to be used to reflect science in general, and astronomy in particular. If such excesses are possible, it is only because we have not known, or have not wanted to know, how to use the possibilities offered by modern means of communication. If astronomers left the field wide open for sellers of gadgets and such-like amateurs to exploit, it was because they did not realise that they ought to play a part in promoting understanding, and that that part had been considerably eased by past cultural changes caused by various apparitions of comets, and particularly by P/Halley. And science gained from the press taking up the event, from people talking about comets, from numerous books having been written, and from the comet having been "sold" in every conceivable fashion. It is a pity that astronomers did not know how to use effectively the tool formed by the media in this last quarter of the 20th century. Let us hope that our descendants in 2061 will be better at it than us; but don't let's fool ourselves: the history of recent apparitions of P/Halley has shown that mankind has difficulty in escaping from its fantasies: comets, and P/Halley in particular, are just one of many. When the public no longer reacts to comets, and apparitions have become just a common-place event for nearly everyone, the wonder will have vanished and knowledge will have reached a limit. We have not yet reached that point, and here in a country where more than one person in three thinks that the Sun revolves around the Earth, selling the comet, fear, and astrology remains a normal part of life. Astronomers will have carried out their duty when it becomes impossible to say just anything, just anywhere, and just when the media like, about objects that astronomers have studied for thousands of years. The "media event" will have become just part of general scientific knowledge, and finally it will be possible to satisfy Arago, who in 1835 said:

"Listen for just a moment to the long addresses made about the future comet and then decide whether we can really be proud of this supposed diffusion of knowledge."

It was Alain, the worthy witness of the 1910 event, who best expressed the change in Man's attitude to the phenomenon:

"Ten centuries ago, as soon as a comet appeared, nearly everyone acted as if they were mad. They expected terrifying events and the collapse of everything. In this they did not think that they were mad, but on the contrary, very sensible. We must remember that for them the regular movement of the stars was the greatest expression of the essential order underlying the world, the apparition of a comet was already a breakdown of that order.

Institutions had to achieve a certain stability and continuity for common sense to acquire its own doctrine, and for the universe gradually to come to be seen as being free from miracles. Consider Halley's Comet and what a number of consistent observations and calculations agreeing with observations were needed to change it from a marvel into a mundane object. Halley, Clairault, Pingré and Pontécoulant, needed not only highly involved methods of calculation, but also the leisure, the security, and the baker at the door every morning. This, friends, is how we all, each in our own way, contribute to building common wisdom, that finally traces the orbits of comets and finally explains the miracle. I can picture a fine new myth: that of Concord chasing away the Gods."

The media clamour that we have just experienced shows that common sense has not yet reached that stage, and that Man still seeks miracles.

A Planetarium for Astronomy

Agnès Acker
Planétarium de Strasbourg, rue de l'Observatoire,
F-67000 Strasbourg, France

Planetariums are continuing to open in Europe and recent years have seen several organizations and meetings devoted to their use. How should a planetarium show be constructed? What are the principles to be followed and the pit-falls to be avoided? This is a problem, because people running planetariums are only rarely astronomers. One great debate centres around the question of whether to use a pre-recorded show or a "live" one.

An automatic show allows a constant level of presentation, however tired or incompetent the lecturer. The Sun rises at the right moment, just as the chosen music starts and either a piece of text or a picture appears. Simultaneous effects can be incorporated, even complicated ones, which often explains a difficult point better.

But it cannot be doubted that the "greatest special effect in a planetarium is the lecturer"! A good planetarium lecturer should simply be a good lecturer, who has a good knowledge of astronomy, who knows how to change their style of presentation according to the audience that is being addressed. A planetarium show should be interactive, the more so, the younger the audience. Questions should arouse interest, and replies should be prepared, which will enable the subjects treated to be understood and memorized. Attention is best captured by direct contact, which is often accompanied by a touch of personal charisma, according to the lecturer's personality.

But such a "dream" lecturer, who has a solid grounding in astronomy is both rare and expensive. Few planetariums are able to obtain one. So I think that the best solution is a pre-recorded programme that incorporates slots for interaction with the audience.

What are the characteristics of a good planetarium show? The aim is to show the universe to members of the general public, without boring them, and without betraying the scientific spirit. A few general principles may be picked out:

1. Always show the sky for that particular season with its essential features: names and properties of the stars and planets, Milky Way, clusters and galaxies visible, etc.
2. Deal with just one theme each time, progressing by stages, using simple and graphic images without being afraid of repetitions. Among the themes to be treated, emphasize "everyday" astronomy (rising and setting of the Sun, seasons, phases of the Moon, red or blue sky, etc.), and topical events (e.g. Comet Halley in 1985–6).
3. Completely avoid mixing inappropriate themes, which causes (or increases) confusion in the minds of the listeners: no parascience, no astrology, no religion. Don't mix "Big Bang" with "Fiat Lux".
4. Avoid reductionism, which over-simplifies discoveries, and tends to lead people

to believe that science in the 20th century can explain everything. After all, we are not yet able to wander around galaxies, and hardly even around the Solar System, contrary to the impression given by some of the media. Only by practising science do we know its limits. An epistemological approach is advisable, enabling one to show the evolution of the sciences over the centuries and the ingenuity of our ancestors, and noting that scientific knowledge is built up layer by layer.

5. Don't emphasize the spectacular aspects to the detriment of the educational ones. Don't attempt to contend with strip-cartoons or science-fiction films, which have no scientific foundation but do have spectacular special effects. In visiting a planetarium, the public is looking for an objective and *faithful* explanation of astronomical knowledge, the "true" universe in contrast to the one shown by astrologers and other charlatans, or by novelists and other audiovisual geniuses. Their work may be very pleasant to read, listen to, or watch, but for them the universe is reduced to a stage set, a falsified backdrop, merely serving to glorify the heroes, the plots, or the stories. In planetariums it is the universe that is the "star", which must not be betrayed with effects that eclipse it (such as laser shows, sound levels that are too high, etc.). Avoid becoming a demagogue (which is not what is wanted) and above all shun any suggestion of "zapping", which is merely superficially, vague, confusing, and counter-productive. Do not use strip-cartoon heroes (Superman or such-like) to get the scientific message across, this only leads to confusion by putting scientific information of the same level as fantasy.

6. Use the planetarium not just as a place of scientific culture, but as one of musical and literary culture also. Express yourself in precise, correct and also elegant language. Don't make any concessions to everyday or slovenly speech; avoid slang. Stud your lecture with quotations from Homer, Bachelard, etc., and with verses from Lamartine, Victor Hugo, Baudelaire, Goethe and others, which will not only enhance the talk but may also help to increase its significance. Use appropriate music and sounds, and silences; draw on the work of Pink Floyd, Grieg, Sibelius, Fauré, Mozart, etc., avoiding too great a dynamic range which will overpower the words.

7. Get to know your public and adapt to it: although scientific knowledge may be objective and universal, the way in which is is spread is rooted in any one culture, and varies according to each country, and each region. Some American planetarium programmes, for example, are studded with local political allusions. Flights of fancy and atmosphere of romance are appreciated by areas of germanic culture.

8. Communicate a taste for astronomy and organize series of talks, training courses, summer schools, etc. for "those who want to know more)e, places where they can meet and work together.

Such a list of imperatives cannot be followed by all professional astronomers, who are too few, and who, perforce, are not always able to speak at a level suitable for the general public. But if a professional astronomer is also a lover of astronomy then they can validly be involved with a planetarium.

Planetariums are being developed in many countries. Whatever their status and size, active cooperation between those running them and local amateurs is obviously necessary and fruitful for both.

Astronomical Publications

Philippe de la Cotardière
14 rue de Castellane, F-75008 Paris, France

The glory of the heavens is available to everyone, and anyone, whatever their level of scientific knowledge, is able to appreciate their beauty, poetry and mystery. They are a window to the inaccessible and the unknown and have always attracted attention and aroused Man's curiosity. This natural fascination for the universe is a precious card in the popularization of astronomy. Three centuries ago, Fontenelle knew how to exploit it in his *Entretiens sur la pluralité des mondes*, which made him famous. Many others followed him in the 18th and 19th centuries, of which Camille Flammarion may be mentioned especially. Printings of his famous *Astronomie Populaire* grossed 130 000 copies between 1879 and 1924.

Science nowadays does not arouse the same enthusiasm as it did in Flammarion's day. Since the development of astrophysics, astronomy has become ever more complex, making the task of those who wish to explain new developments more difficult.

It seems, however, that the public's curiosity about astronomy has never been as great: everywhere you find lecture halls full, observatories swollen with visitors, planetariums sold out. Undoubtedly, this does in part arise from the interest that I mentioned at the beginning, but it is also caused by the media's response to the exploration of the Solar System in the last twenty years by spacecraft, and to the the spectacular discoveries that they have made. Finally, it perhaps also stems from the general availability of amateur-sized instruments, thanks to which anyone can now enjoy the delights of observing the sky with a refractor or reflector at a modest price.

This public interest in astronomy has not escaped publishers. Some recent popular works have had record printings and at long last we can see, in some bookshops, a range of astronomy books that rivals the number on astrology. The books available fall into approximately three categories: encyclopedic works, heavily illustrated and showing a general view of the universe; smaller books, moderately or much less profusely illustrated, that describe recent developments in a specific field such as cosmology; and practical books (observation of the sky and instrumentation), at various levels.

A detailed examination of books in each group shows that certain aspects of astronomy are preferred, because they have been exploited more by the media. Recent discoveries about the Solar System or modern views of cosmology with their description of the Big Bang, undoubtedly benefit from favourable treatment. Astrometry, on the other hand, remains completely overshadowed.

We must also deplore the absence of elementary works designed specifically for young people. Works for children are, in general, picture books where the text has just the role of providing an explanation of the pictures. One of the areas in which popularization by means of books ought to concentrate, is that of documentary books for children, all the more as such a well-executed book is also bound to be read by their parents.

The easy solution, for a publisher, is to buy rights to a book that has sold well in other countries, translating it, or perhaps adapting it. Until recently this was, unfortunately, the most common course in France, sometimes with dreadful results (with numerous errors in translation). But now we are seeing a very definite movement towards works that originate here. For practical works, associations representing amateur astronomers or some of their members, known for their work, are the sources favoured by publishers, who in effect know that the associations are useful contacts able to promote such works.

For the encyclopedic works that require various contributors and for popular works on some specific theme, the choice of authors is paramount, but often delicate. Some eminent scientists may prove to be talented popularizers: Eddington and Gamow were, for instance. Others have difficulty in getting their message across, the problem being not so much that of being simple as of being inspiring: readers of a popular book always need to be stimulated by what they are reading. My personal experience leads me to state that the best results may arise from close collaboration between a researcher and a scientific journalist. Scientists are often tempted to pay excessive attention to basics and to neglect the overall impression, because, good intentions notwithstanding, they are thinking of the reaction of their peers, and not that of their readers. This is one of the problems that collaboration between a researcher and a journalist can avoid.

Whatever happens, a popular work will not be a success unless it is written with enthusiasm. If the author is bored by writing a book, then the reader will be even more bored by reading it. Remember Camille Flammarion, whose enthusiasm is quite obvious on each page of his work.

Illustrations are an important part of modern, popular-astronomy books. There are rich resources for Solar-System objects, thanks to the many photographs obtained by spaceprobes, but for galactic or extragalactic objects, on the other hand, the same pictures are found in one book after another. Luckily an injection of new life is occurring with photographs from major observatories in the southern hemisphere. Artwork too, often plays an important part in understanding the text and considerable effort has been given to this aspect in many recent books.

Apart from books, the interest that astronomy has for a wide public, avid to hear about the latest information, explains the presence of articles about recent developments in astronomy that can be found in must issued of news or popular-science magazines at all levels.

Unfortunately, magazines specifically devoted to the popularization of astronomy mostly suffer from being the journals of organisations that are non-commercial, and which neither have the technical expertise nor the financial resources to ensure a wide circulation, although the journals themselves provide information of the highest standards.

In default of being able, without external help, in issuing publications that have a wide circulation, amateur associations can make regular contributions to media that do reach a wide public. Items of the "month's sky" type can find a place in the press, or even on television. Some large newspapers already carry this sort of item, either in collaboration with individual astronomers or associations. Developments in amateur astronomy can only prompt the editors of newspapers to such ideas. Conversely, regular publications of this sort greatly help to popularize astronomy.

Other links can be forged with publications that reach readers to whom astronomy might be a source of interest. Avid photographers, computer buffs and handymen all have their own magazines, and each of these could lead to ideas for articles able to contribute to the popularization of astronomy.

Popularization of astronomy in books and magazines is a never-ending task, given the constant progress in knowledge, and all the enigmas that arise new discoveries. Current events are a distorting mirror that often lead to emphasis on minor subjects. But astronomical events also allow one to get back to basics. This is all the more necessary when we recall the results of a survey made several years ago by the Centre National d'Etudes Spatiales, which showed that one French person in three over fifteen still believes that the Sun moves round the Earth.

Problems in Publishing a Popular Journal in a Small Country

Pierre Gugnon and *Jacques Sauval*
Société Royale Belge d'Astronomie, Observatoire Royal de Belgique,
Avenue circulaire, 3, B-1180 Brussels, Belgium

Every journal with a small circulation has the same problems and *Ciel et Terre* is no exception. The basic cause is the same everywhere: the first thousand copies is far more expensive than the second and subsequent thousands by a factor of 6 to 8 times. This obviously arises from the cost of origination of material. All low-circulations journals have the same financial problems.

We have found that it is beneficial to periodically review the following points: the printing process, the format, the frequency, the editorial team, distribution, promotion, subscription fee, etc. In recent years we have found this process to lead to considerable savings.

Photolitho printing in A4 size has considerable advantages, and bi-monthly publication balances costs of postage against size of journal and reader satisfaction. Our journal is distributed to all subscribers and special numbers are sold by specialist booksellers and other astronomical groups, and sent free of charge to daily and weekly newspapers with a request that it be mentioned (with very good results).

Supplementary sources of income arise from special issues, such as annual ephemerides, and publications on the Sun, comets, etc. which can interest a wide public if they are properly promoted and if the selling cost is fixed low. This can be achieved because they are normal bi-monthly issues, so the additional numbers can be printed at relatively low costs. Advertisements are a useful but moderate source of income.

Other economies can be made by setting the text direct from disk, providing diagrams to final size, etc.

We would suggest that there could well be assistance between the various journals, perhaps even with exchange of material. If the editors of any astronomical journals wish to discuss any of the ideas that we raise, we shall be pleased to help.

Methods of Popularizing Astronomy in Various Countries

Cecylia Iwaniszewska
Institute of Astronomy, N. Copernicus University,
PL-87-100 Torun, Poland

Abstract

Some interesting methods of popularizing astronomy in various countries were reviewed, and may be divided into two types: active and passive methods. Active methods include:

1. Astronomy in the countryside. Astronomical camps for both children and grown-ups, who normally live in towns, to learn basic astronomical facts. (Italy)
2. Astro-puppets. Visitors to an observatory are greeted by puppets (Copernicus, Galileo, etc.) who not only give talks, but also converse with the audience. (Argentina)
3. Hand-operated devices. A simple orrery showing the movement of the Earth, and other devices that viewers operate for themselves. (India)
4. Graphical calendars. Several popular observatories collaborate to produce a yearly calendar showing planetary rising and setting times, etc. (Czechoslovakia)
5. Amateur clubs. High-school and university students became so interested in astronomy that they formed their own group and now produce a magazine and carry out observing, etc. in a country where there is no professional astronomical institution. (Paraguay)
6. An interdisciplinary approach. Various meetings and workshops are arranged to bring together astronomy, physics, biology, geology, etc., leading to a better understanding of modern science. (Japan)
7. Astronomical competitions. Everything from children's drawings to special tests and papers, sometimes connected with special events. Also awards for the best work of popularization in a given year. (Worldwide)

Passive methods are far more traditional, and include planetarium shows, including special ones for pilgrims (India); special exhibitions; radio and television programmes; public lectures; popular magazines usually edited for special groups of readers. Finally, what about humorous astronomical stories or pictures, is that active or passive?

[A large number of contributions described various means of popularizing and teaching astronomy. As many relied heavily on illustrations that cannot be reproduced here, or on practical demonstrations, these contributions are covered by the following report, which emphasizes the salient points in each paper.

Two contributions dealt with general questions affecting popularization of astronomy, and they were accompanied by four contributions describing different, practical methods of conveying information to various audiences, including a special campaign in a a rather unlikely site – the Paris Métro. – Eds.]

Opening the Sky to Everyone

Joel Chanoir
CEMEA, 43 bis rue des Moissons, F-51100 Rheims, France

CEMEA [Centres d'Entraînement aux Méthodes d'Education Active – Centres for Training Methods of Active Education] exists to encourage better understanding of science among teachers, organisers of youth groups, leaders of summer camps, etc. Its activities with regard to astronomy were described.

[Text not received – Eds.]

Interesting Business in Astronomy

Michel Lyonnet du Moutier
23 av. des Bleuets, F-92700 Colombes, France

Support from firms is of increasing importance in amateur astronomy. Two projects were described. In the first, a company running a winter resort in Argentina was persuaded to hold Halley observation sessions at their site in the Cordillera de los Andes. A similar event was later held by the same company at a new resort in southern France. The second project involved interesting a professional journal for construction engineers in producing a special issue devoted to Comet Halley. This was speedily exhausted, most of the demand for back numbers coming not from the engineers themselves but from their families.

Preparing Astronomical Videotapes

Françoise Suagher[1], *Pierre Chopard*[1], *Pierre Magnien*[1], *Jean-Pierre Marchand*[1], and *Jean Paul Parisot*[2]
[1] Association Astronomique de Franche Comté, 34 av. de l'Observatoire, F-25000 Besançon, France
[2]Observatoire de Besançon, 41 bis av. de l'Observatoire, F-25044 Besançon Cedex, France

In 1982, the Association Astronomique de Franche Comté prepared a slide show entitled "Meet History's Comet". This had been so successful, being presented in more than 100 towns, that a second show entitled "Worlds in the Sky" about the Solar System had been prepared and made by an audio-visual workshop. The high cost (20 000 Fr) had been largely covered by a grant.

Transmission of Astronomical Data and Information by the A.F.A. Videotext System

Bernard Pellequer
Observatoire Astronomique, B.P. 14, F-34150 Aniane, France

Following the introduction of a videotext system in France using MINITEL terminals, the Association Française d'Astronomie organized an astronomical information service. This covers celestial and space events, location of the planets, lists of societies and clubs, bibliography, etc. A practical demonstration of the system in use was given.

The Solar System Promenade

Andreas Tarnutzer
Schweitzerische Astronomische Gesellschaft, Hirtenhofstrasse 9, CH-6005 Luzern, Switzerland

In order to convey the dimensions and relative scale of the Solar System all sizes and distances are reduced by a factor of 10^9, resulting in a solar diameter of 1.4 m, a diameter of 5 mm for Mercury, and a mean Sun–Pluto distance of close to 6 km. Stainless steel models of the planets (and the Moon) are mounted at appropriate distances from a yellow steel sphere representing the Sun. Ten of these Solar System Promenades have been constructed in Switzerland.

The Open Sky Underground

Igor Beauvois and *Olivier Las Vergnas*
210 av. Pierre-Brossolette, F-92240 Malakoff, France

From 1985 March 26 to April 5, a special public astronomical education campaign was mounted in 16 stations of the Paris Métro. Apart from static displays, items included two planetariums, mirror-making, numerous round-table discussions, astronomical films, etc. Millions of people saw some part of these displays and surveys showed that many returned several times. The event was coordinated by the Association Nationale Science Techniques Jeunesse [National Association for Youth Science and Technology] and involved many official and private bodies, 30 astronomical groups, and 420 employees of the RATP [Régie autonome des transport parisiens – Paris city transport authority].

[Four contributions dealt directly with the question of the teaching of astronomy in schools. This whole subject is the topic of the forthcoming IAU Colloquium 105 "Teaching of Astronomy: Present and Future". – Eds.]

Teaching Astronomy in Schools

Cecylia Iwaniszewska
Institute of Astronomy, N. Copernicus University,
PL-87-100 Torun, Poland

This contribution summarized the ways in which astronomical teaching was evolving and some of the teaching aids that could be employed. The work of GIREP (Groupe International de Recherche sur l'Enseignment de la Physique) – which holds conferences on the teaching of physics, including astronomy, every two years – was described together with some material reported at the 1986 conference which was devoted to astronomy.

Teaching Astronomy and the Vital Role of the True Amateur

H. Robert Mills
83 Firs Road, Firsdown, Salisbury, Wilts. SP5 1SW, U.K.

This contribution summarized the situation in the United Kingdom where there are 5 organisations that are concerned with astronomical observation, but where there are hopes that a single unifying committee may be formed. Practical work for pupils was essential and many examples of simple equipment and practical applications were shown.

Astronomical Education in Poland

Katarzyna Turaj
Institute of Physics, Wyzsza Szkola Pedagogiozna,
ul. Podchorazyc nr. 2, PL-30084 Krakow, Poland

Astronomical education is undergoing a fundamental change in Poland with the introduction in 1987 September of a revised astronomy programme. Although previously taught as a separate subject, it was often neglected by pupils concentrating on examinations for other subjects. Astronomy is now treated as part of the physics course and will be taught to all grades of pupils. This contribution also described the encouragement of astronomy through school astronomy groups, summer camps and the Astronomy Olympiads, where the winners are excused from entrance examinations to university if taking astronomy or physics courses.

Astronomy and Computing in Spanish (Aragon) High Schools

Antonio Elipe[1], *Joaquín Ezpeleta*[2], and *Víctor Lanchares*[1]
[1] Dpto. Astronomía, Universidad de Zaragoza, E-50009 Zaragoza, Spain
[2] Dpto. de Informatica, Escuela de Magisterio,
E-Pamplona, Navarra, Spain

Astronomy is an optional subject available at some schools, but is generally underfunded. A recent government campaign aims to increase the knowledge of computing and astronomy has been fruitfully linked with this. Details of three specific areas were given: computer-based training; modelling and simulation; and guided programming, where the pupil has to master both astronomical and computational problems.

[In addition to the paper by Dr Acker (given above), four contributions dealt with planetariums and their programmes. These formed a sharp, and interesting contrast, because their size and facilities differed so widely. Further information about European planetarium facilities may be found in "European Planetarium News", published by the Planetarium at Strasbourg. Two Colloquia on European Planetariums have been held in Strasbourg, the first in 1984 and the second in 1988. – Eds.]

The Planetarium at the Palais de la Découverte

Gérard Oudenot
Départment d'Astronomie, Palais de la Découverte,
av. F.D. Roosevelt, F-75008 Paris, France

This large planetarium has about 250 000 visitors every year. It seats 201 persons and has a modern Carl-Zeiss Jena projector. Two or five showings from a series of about ten different programmes are given daily to the general public. Special programmes at a much higher level are presented in conjunction with navigation, astronomy and other academic courses.

The Starlab Mobile Planetarium in the Mid-Pyrenees

Jean-Pierre Chrétien, Jean-Pierre Brunet
1 allée des Roses, F-31520 Ramonville Saint-Agne, France

This planetarium with inflatable dome is available to schools, colleges and youth centres in the region. Each institution using the equipment provides its own lecturer, supported by various levels of documentation, discussion sessions and alternative displays. In the first year about 50 organisations used it, each for about a week, with some 15 000 child, and 6000 adult visitors. No serious problems were encountered despite this intensive use.

The Planetarium at a School in Nice

Jeanine Chappelet
Planetarium du collège Valéri, B.P. 86, F-06013 Nice Cedex, France

An observatory, laboratory and planetarium were set up by pupils at a school in Nice with assistance from the Club des Pléiades. Various presentations were given, for children and adults, and some with the assistance of astronomers from the Observatory at Nice for teacher-training.

[Full text not received – Eds.]

The Planetarium as a Medium for Teaching Astronomy

Kunda Barve
Planetarium, Nehru Centre, Dr Annie Besant Road, Worli,
Bombay 400018, India

Being sited in a large city (with about 10 million inhabitants), the planetarium at Bombay has to try to meet the requirements of a very wide range of visitors, some 3.3 million of whom have seen presentations. These visitors come from very different linguistic, cultural, religious, economic and educational backgrounds, which pose many problems in arranging suitable programmes. (There are 16 major languages and many less common ones.) School children make up the bulk of visitors and presentations for them are linked to their academic grade. Amateur astronomical organisations are gradually coming into existence in India, and the planetarium provides help and encouragement to them. Special shows are made for many individual groups of people, ranging from lovers of poetry and fine arts to invalids and the handicapped. A wide range of additional activities such as competitions, field trips and exhibitions is also provided.

[The activities of amateur societies and groups, including coordination on a national scale, were covered in five contributions. – Eds.]

Twenty Years of the Japan Amateur Astronomers' Convention

Seiji Kimura
Kita Otsuka 2-33-19, Toshima-ku, Tokyo, Japan

In 1967, the Kawasaki Astronomy Club organized the first national assembly for amateur astronomers. This was such a success that a second was held in 1969 and yearly thereafter. Since the establishment in 1982 of the Kaho Prize commemorating Kaho, who was the first Japanese to discover a new comet (in 1936), this award has been presented at the Convention.

Astronomy in the Big Apple

John Pazmino
Amateur Astronomers' Association, 1010 Park Avenue,
New York, NY 10028, U.S.A.

New York is probably unique in metropolitan cities worldwide in that it has no "official" or "public" astronomical institution. The Amateur Astronomers' Association, an amateur body, fulfils that role and answers many queries from the general public and various organisations. Founded in 1927, it has around 500 members. Apart from providing all the usual services to its members, it also organizes regular observing sessions in parks and open spaces away from the worst of the light-pollution. It staged very successful public Halley watches in which over 10 000 visitors participated.

The Activities of Astronomical Clubs

Christian Bourdeille
5 av. Carnot, F-77220 Gretz-Armainvilliers, France

Recent studies have shown that there are some 60 000 amateur astronomers in France. This has been reflected in the number of astronomical clubs that have been formed. This contribution summarized the aims and activities of such clubs, the conditions for their success, and their contribution to scientific astronomy.

The Flammarion Astronomical Society of Montpellier

Jean-Michel Faidit
24 av. de la République, F-30190 St. Chaptes, France

The Flammarion Astronomical Society of Montpellier was founded in 1902 and remained very active until 1923, when the death of certain prominent members, precipitated its own demise. An observatory in the Tour de la Babote was opened in 1903 but fell into disrepair in the 1920s. One prominent member was Marcel Moye, who on his death in 1939, left his instruments to the S.A.F., accompanied by the sum of 100 000 Fr, half of which was used to endow an award for young astronomers.

A Decade of Organized Work in Venezuela

Domingo Sanchez
Sociedad de Astronomía Guri, Apartado 368, Puerto Ordaz 8015-A, Venezuela

In 1978 Venezuelan amateur astronomers decided to meet yearly to provide a forum for the exchange of ideas and discussion of observing programmes. The contribution summarized the ten years of work and the papers that had been presented to the meetings.
[Text not received – Eds.]

[Public observatories and the use of professional facilities by amateurs were described by four participants. – Eds.]

The Bochum Observatory

J.V. Feitzinger, M. Hünerbein, R. Kordecki, G. Monstadt, and *J. Prölss*
Ruhr-Universität Bochum, Astronomisches Institut, Postfach 102148,
D-4630 Bochum 1, Fed. Rep. of Germany

The various units of this observatory consist of a planetarium – which has a complex computer-controlled system – the programmes for which are varied with different degrees of astronomical information and entertainment value; observing stations with 12.5, 40, and 60–cm instruments, used for regular public observing nights, special events, and special programmes for young children and their parents as well as specific older groups; a mobile observatory with a 32–cm telescope; and a series of astronomical courses.

The Triel Observatory

Jean-Paul Trachier
Observatoire de Triel, Groupe Astronomique des Yvelines,
F-78510 Triel, France

Although with limited equipment, this small observatory mounts a permanent exhibition and conducts various astronomical courses. It is primarily concerned with astronomical education and is visited by 5000–6000 pupils every year.

[Text not received – Eds.]

The Bruno-H.-Bürgel Observatory

Helmut Busch
Bruno-H.-Bürgel Sternwarte, DDR-7302 Hartha (Kr. Dobeln), GDR

The observatory is both a school and public observatory. Lectures are arranged for pupils at various levels as well as courses for teachers. Since 1959 astronomy has been a compulsory subject in schools in the D.D.R. (one of only four countries in the world in which this is the case). The observatory also participates in scientific work, particularly in the field of variable stars, where it has always had very close links with Sonneberg Observatory.

[The English-language text of this contribution is published in *Circular 67* of the Variable Star Section of the British Astronomical Association. – Eds.]

The Strasbourg Observatory

Roger Hellot
17 rue de la Ziegelau, F-67100 Strasbourg, France

The Alsace Group of the S.A.F. is closely associated with the Strasbourg Observatory and is responsible for supervising visits by the public after planetarium shows. The 6-inch refractor was restored by three of the amateurs and the Group has the use of the 21-cm refractor for planetary photography and other projects.

[With the increasing public interest in astronomy, professional astronomical institutions have devised many ways of handling enquiries and also carrying out general public education. Specific details of some of the methods used were given in four contributions. – Eds.]

Feeding the Five Thousand

Muriel Enock
1230 Verdier Avenue 102, Brentwood Bay,
British Columbia, Canada V0S 1A0

The Dominion Astrophysical arranged 20 special viewing sessions for Comet Halley, and although most of these were clouded out, some 3000 people attended. The Observatory receives 30 000 visitors annually and its display area was recently renewed. This was done on a very limited budget, but the display succeeded in providing both scale and distance models of Solar-System bodies, telescope models, and much visual material, all accompanied by bi-lingual text.

Efforts by the Instituto de Astrophisica de Canarias to Publicise Astronomy

Ignacio Garcia de la Rosa, J.J. Cepa, M.R. Kidger, A. Mampaso, L. Martinez, F. del Puerto, C. Sanchez, Ll. Tomas
Instituto de Astrophisica de Canarias, E-38200 La Laguna (Tenerife), Spain

The Institute provides facilities for school and group visits and these have also included "Astronomical Weeks" (held on more than one of the islands in the group) and yearly open days. In 1985 the formal opening of the Institute engendered so much publicity that at the first open day at the Rocque de los Muchachos Observatory, 15 000 people attended. (The island's total population is 80 000.) During 1986, a special event organized at a beach on Tenerife was attended by an estimated 70 000 people, thousands more being prevented by an enormous traffic jam. A month later, the inhabitants of La Palma switched off all private and street lights throughout the island for a period of 3 hours. Because of the interest and demand from local people and tourists, additional visitors' centres are planned.

Tailor-Made Messages: Astronomical Information for Diverse Audiences

James Cornell
Smithsonian Astrophysical Observatory, Cambridge, MA 02138, U.S.A.

The SAO offers a wide range of public-relations services. For example, research results may be transmitted, via reporters, to newspapers or television stations. More direct approaches include "Observatory Nights", special events for children, series of popular lectures, and publication of lecture series as popular books. Television, radio and film are used whenever possible to provide information about general and specific events. Standard information packages answer many of the 5000 written requests received annually. Day-to-day queries are handled (in part) by an automatic telephone answering service. At the SAO's Whipple Observatory in Arizona (site of the MMT), visits are by guided bus tours from the foot of the mountain 18 miles (32.5 km) away. Concrete pads have been installed at the base-camp for the use of amateur astronomers – perhaps the only amateur observing site connected with a major U.S. observatory.

The Role of the Observatoire de Paris, Meudon in Providing Astronomical Information

M. Floquet and *Brigitte Schmieder*
Observatoire de Paris, F-92195 Meudon Principal Cedex, France

The three research centres (Paris, Meudon, Nancay) receive groups totalling about 10000 visitors per year, consisting primarily of school-children and the general public. Tours begin with a slide-show and then visits to the instruments are guided by an astronomer. One-day courses for teachers are held at Meudon and Nancay once a month and observing evenings at Meudon. Open days are held frequently, and in May 1985, despite dreadful weather, 30000 people visited Meudon. Each centre has prepared brochures describing its work, together with lists of postcards, slides and other material. Three slide-shows are available, and others are in preparation.

Observations – Concluding Remarks

Brian G. Marsden
Harvard-Smithsonian Center for Astrophysics, 60 Garden Street,
Cambridge, MA 02138, U.S.A.

I must speak for all the professional astronomers here in saying that I am most impressed by the contributions made at this IAU Colloquium by amateur astronomers. I spoke earlier of the contacts between amateur astronomers and the IAU Central Bureau for Astronomical Telegrams and Minor Planet Center, particularly with regard to discoveries of comets, novae, supernovae and minor planets. But I also spoke of the great importance of follow-up observations and computations. Janet Mattei has quoted Fran Cordoba's remark that "discoveries belong to those who keep vigil on them". It is difficult to deny that the routine of follow-up lacks the glamor associated with making a spectacular discovery, but such discoveries are few and far between, and we need to give greater credit to the unsung heroes who go steadily about this work. Even if all the professional astronomers in the world wanted to pursue this routine, there would not be enough of them to monitor everything that needs to be watched. This is precisely where the amateur comes in.

Although observational work by amateur astronomers is nowadays conducted in a variety of ways, the predominant mode of observation is still by the traditional use of telescopes visually. Visual work discussed at the conference involves (a) discoveries (of comets, novae, supernovae and other variable stars); (b) magnitude estimates (of the same objects); (c) observations of Jupiter's satellite phenomena (including mutual events); (d) identifications of double stars (for HIPPARCOS); (e) counts of sunspots; (f) micrometric measurements of sunspots and studies of sunspot motions; (g) inspection of transient features on the planets); and (h) naked-eye observations of meteors (for the identification of new showers and the evolution of established ones).

Photographic work conducted by amateurs concerns (a) discoveries (also now of minor planets); (b) recoveries of comets and minor planets; (c) astrometry (of comets, minor planets and double stars, some of the minor-planet work being in connection with occultations or near occultations of stars); (d) investigations of cometary features (near-nucleus activity, tail disconnection events); (e) observations of sunspots; (f) measurements of isophotes at solar eclipses; (g) double-station detections of meteors; and (h) observations of deep-sky features such as galactic bridges.

Photoelectric observations are becoming increasingly popular with amateurs, principally of course for photometry of variable stars, but some amateurs are also doing work in polarimetry.

Spectroscopic work by amateurs tends to need a good strong light source, so it is reasonable that much of it should involve studies of the sun in H-α light or the measurement of radial velocities in solar flares. Nevertheless, we have also heard of stellar slitless spectroscopy down to magnitude 10–11. Spectroscopy is clearly an area worthy of further attention by amateurs in the future, particularly as it pertains to transient or otherwise variable celestial objects.

The astronomical use of video is relatively new, but amateurs have been quick to adopt it. Video is used most extensively in occultation work, but it also plays a role in the study of solar eclipses (including observations of Baily's beads and measurements of the solar radius). It has also been used to detect the rotation of artificial satellites and in scintillation studies.

Other observational activity by amateurs has included CCD imaging of galaxies, balloon detections of micrometeors, and radio observations of meteors and the sun.

Amateurs are also very much involved with astronomical computing, and since much of this computing is closely related to observation, it is appropriate to mention it here. Orbit computation is a topic to which amateurs make particularly useful contributions. This applies to comets and minor planets, and also to double stars. There are also extensive analyses of lightcurves of variable stars and comets, with results ranging from determining parameters of eclipsing-binary systems to examining problems of thermal equilibrium. Amateurs are concerned with computerized databases and have investigated the cometary dust production required to match observed cometary dust tails.

Most of the above research is conducted by amateur astronomers using their own equipment. Impressive work is also done by amateurs using professional equipment, sometimes – but not always – in collaboration with professional astronomers. Amateurs use professional equipment most notably at Pic du Midi (a 60-cm reflector and a 50-cm refractor), but also at Strasbourg and Meudon. Mention can also be made of the proposed involvement of amateurs in programs to be conducted with the Hubble Space Telescope. In addition to telescopes, measuring engines are items of professional equipment that can be and have been particularly utilized by amateurs.

Ennio Poretti remarked that professional astronomers must drop their prejudice against amateurs. At the same time, however, it is necessary for amateurs to demonstrate the reliability of their work. Specifically, Zdenek Kviz pointed out that amateurs should be sure to give enough information so that future researchers can assess this reliability. As an example of the kind of problem that can occur, let me mention the episode of the Aries (or Perseus) flasher. The amateur astronomer principally involved in this episode first telephoned me about these strange observations toward the end of 1984. The only opportunity I was given to assess his reliability was his own statement that he was the leading amateur meteor observer in his country. I refused to publish the observations in the IAU Circulars. A few months afterward I was startled to see them mentioned in Sky and Telescope. Later I was later absolutely appalled to see them discussed at length in a paper in the Astrophysical Journal Letters. This paper, written with several collaborators, some of them professional astronomers, had been subjected to the normal refereeing process of this professional publication, and it was finally accepted on the grounds that the editor did not want to be accused of rejecting an intriguing result just in case in might conceivably be valid. But it was obvious to numerous astronomers, amateur and professional, that the data being discussed were most unsatisfactory. Not only did the authors not demonstrate their reliability, but the bias of the professional editor toward amateurs took the very worst form – that of patronization.

I have occasionally been accused of an anti-amateur bias, in that I have not accepted results submitted by amateurs for publication in the IAU Circulars or Minor Planet Circulars. This is unjust: professional results have been rejected too, and I have merely insisted that reports by amateurs and professionals meet the same standards. One of the amateur astronomers whose work I have rejected has complained that he is "only an amateur and does not have access to the facilities and the knowledge of professionals". This is not an acceptable excuse. If an amateur expects recognition in an area of astronomy that is also practised by professionals, he must play by the same rules. In responding to the Aries flasher paper in the *Ap. J. Lett.*, Paul Maley, an amateur astronomer with a great deal of experience, demonstrated that the principal evidence put forward concerning the flasher – a single photograph – was in all probability the observation of a glint from the artificial satellite Cosmos 1400. Maley is an amateur, but his demonstration was conducted in a completely professional manner.

So, when it comes to discussing specific astronomical observations, let us not distinguish between amateur and professional. If we are to make any distinctions, they should surely be only on the grounds of competence.

International Astronomical Union – Colloquium 98 General Session

Concluding Remarks

André Koeckelenbergh
Vice President, Société Royale Belge d'Astronomie,
Méteorologie et Physique du globe, Observatoire Royale de Belgique
Avenue circulaire 3, B-1180 Brussels, Belgium

My habit of visiting bookshops has led to my discovering this morning an example of the major work by the philosopher Helvetius, *De l'esprit* [Concerning the mind]. In chapter IV, "The abuse of words", I found this statement: "Descartes had already said, before Locke, that peripatetic scholars who hid behind the obscurities of words are very similar to blind men, who, in order to make their fight more equal, pull a sighted man into a dark cave. If a man knows," he said "how to give precise ideas to the words that he uses, then his triumph is assured." This is the prime virtue of this discipline that we study and is the cause of the shining example that it presents.

The practical conclusion that emerges from the papers given on the subject of "The history of amateur astronomy" is that this field is a veritable mine of inspiration for historical work of considerable interest. I would add that their scientific usefulness is not diminished by the necessity of having to find, in the past, the correct assessment of factors that actually vary secularly.

Ever since the 17th century, men of science have carefully recorded their observations in tomes that are now gathering dust in the archives. This is an immense field in which we can exercise our curiosity. Who better than astronomers, used to the thousand pitfalls that lie in practical observation, than to decipher ancient tomes and to interpret the observations that they contain?

Although it may be difficult to define what we mean by an amateur astronomer, it may nonetheless be possible to provide a definition that sets some limits on our interpretation. An "amateur" is any lover of astronomy who does not make it their job or their main source of income. We have been given distinguished examples of authentic amateurs of the past, as of the present, who have been converted, in either a permanent or a temporary fashion, into professionals. How many of our colleagues, specialists in one field, are respected amateurs in another? How many young graduates, either having reached the end of their studies or the end of a research grant, find themselves back on the jobs' market and relegated to the role of enlightened amateurs. They bring to societies and clubs the fruits of their knowledge of the most modern technical advances.

On behalf of Professor Etienne Bernard, the current president of the Société Royale Belge d'Astronomie, de Meteorologie et de Physique du globe, and also as an already veteran member of the Société Astronomique de France, allow me to make this grateful tribute to it by its "little Belgian sister".

It is thanks to Camille Flammarion and his *Astronomie Populaire* that amateur astronomy expanded in Belgium. Our libraries at home also included work by Amédé Guillemin, Charles Nordman, Théophile Moreux, Lucien Rudeau, Pierre Rousseau and Paul Couderc. But the Belgian public's interest in astronomy goes even farther back.

In 1823, Lambert Jacques Adolphe Quetelet, who was the founder and the first director of our Royal Observatory, created a successful "public and free" course of lectures in astronomy, which I have the privilege of continuing today, one hundred and sixty-fours years after it began. In 1827, Quetelet gathered the texts of his lessons into three small volumes entitled *Astronomie Populaire*, following the example set by the one edited by his friend and colleague, François Arago. The friendship that linked Quetelet, Arago and Alexis Bouvard was deep, and founded on both respect and fellow-feeling. In his *Souvenirs de ma jeunesse* [Memories of my Childhood], recently published again, Arago describes this with warmth and considerable emotion.

Around 1845, the group around Quetelet were joined by a young amateur from Mons, who was to prove quite extraordinary: Jean Charles Houzeau de Lehaye. He became an astronomical assistant, a workman, an international journalist, an explorer, an adventurer, a free-thinker, an abolitionist, a follower of Fourier[1], a socialist and a republican. It was he whom King Leopold II chose, disregarding his ministers' reservations, as successor to Quetelet. He was to revitalise the Royal Observatory.

Under the inspired leadership of Houzeau, and with help from collaborators, the journal *Ciel et Terre* was published in 1880, and has continued ever since. *Ciel et Terre* preceeds your *Astronomie*, which was born in 1882. The Sociéte Astronomique d'Anvers was founded in 1893 and the Sociéte Belge d'Astronomie itself was set up in 1895.

Currently, there are societies and local clubs, most particularly in the French-speaking part of the country, in Liège, Brussels, Tournai, Dinant, Charleroi, as well as the Apex group. Since 1940, the Vereeniging voor Sterrekunde, Meteorolgie en Geofysica caters for the Flemish-speaking area.

All of us feel that we are descendants of Camille Flammarion.

Where necessity knows no law, amateurs and professionals are perfectly able to work together in complementary ways. In closing the session devoted to the history of amateur astronomy, I praised the people who take part in expeditions to observe solar eclipses and also those who contribute to the sunspot observing network. But there are also the indefatigable variable-star observers, meteor watchers, the talented planetary observers, and all those who discretely infiltrate professional research groups. Others I have not mentioned, but all deserve to be praised for both the vigour and for the seriousness that they display.

Such activities help to bring both the knowledge and the practice of scientific work into everyday life. In this way we are contributing towards the elimination in

[1] Fourier, Fançois Marie Charles (1772–1837), a French social reformer, who advocated the abolition of many social conventions and the establishment of communities (or *phalanges*) of 1500 to 1800 persons, with appropriate individual and communal labour. – Eds.

the minds of the public those rather superstitious fears about science and progress that we encounter everywhere, and which are exploited by irresponsible people. This is how we can more effectively fight the resurgence of pseudo-science and ensure, as Goya said, that "the sleep of reason" does not "give birth" to yet more "monsters".

List of Participants

AFRIQUE DU SUD / SOUTH AFRICA

José A. da SILVA CAMPOS (Durban)
Jan HERS (Sedgefield)
Michiel OVERBEEK (Evendale)

ALGERIE / ALGERIA

Benameur BELGRAINET (Saïda)
Madjoudj NAWEL (Kouba)

ALLEMAGNE (Rép. Fédérale d') / Federal Republic of GERMANY

Wolfgang BEISKER
Dr. H.J. BODE (I.O.T.A./E.S., Hannover)
Dr. Werner CELNIK (Bochum Observatory)
Livia CORDIS (Berlin)
Prof. Dr. Johannes FEITZINGER (Bochum Observatory)
Stefan HAACKE (Gières, France)
Dr. Andreas HAENEL (Osnabrück Planetarium)
Guido HESTERMEYER (Wallenhorst)
Robert HILZ (Berlin)
Dieter MAIWALD (Berlin)
Hans-Ludwig NEUMANN (Offenbach)
Klaus REINSCH (Berlin)
Michael STEPHAN (Berlin)
Carol WITZKI (Bochum)

AUSTRALIE / AUSTRALIA

Dr. Zdenek KVIZ (Kensington, NSW)

BELGIQUE / BELGIUM

Roland BONINSEGNA (Dourbes)
G. CANONACO (Genk)
Léo HOUZIAUX (Liège)
J. GIJSENBERG (Genk)
Jan JANSSENS (Mortsel)
Dr. André KOECKELENBERGH (Observatoire Royal, Bruxelles)
Dr. Jacques SAUVAL (Observatoire Royal, Bruxelles)

CANADA

Allan COOK (Portage La Prairie, Manitoba)
Muriel ENOCK (Brentwood Bay, Brit. Col.)

DANEMARK / DENMARK

Dr. Niels WIETH/KNUDSEN (Frederiksberg)
Mme Niels WIETH/KNUDSEN "

ESPAGNE / SPAIN

J. Antonio BERNEDO (Madrid)
Jaime BUSQUETS (Valencia)
Antonio ELIPE (Zaragoza)
James ELLERBE (Nerja, Malaga)
Amando ESTELA (Valencia)
Joaquin EZPELETA (Pamplona)
Juan FABREGAT (Valencia)
Dr. Ignacio GARCIA de la ROSA (La Laguna, Tenerife)
Angel GOMEZ-ROLDAN (Madrid)
Victor LANCHARES (Universidad de Zaragoza)
José RIPERO OSORIO (Madrid)
Diego RODRIGUEZ (Madrid)
David VALLS-GABAUD (Le Vésinet, France)
Enrique VELASCO (Madrid)
Pedro VELASCO (Madrid)

ETATS-UNIS / UNITED STATES

Dr. James CORNELL (Cambridge Observatory, Mass.)
Dr. Stephen EDBERG (JPL, Pasadena, Calif.)
Richard EVELYN (Marietta, Georgia)
Charles HEROLD
Paul MALEY (I.O.T.A., Houston, Texas)
Dr. Brian MARSDEN (Cambridge Observatory, Mass.)
Dr. Janet MATTEI (AAVSO, Cambridge, Mass.)
Ben MAYER (Los Angeles, Calif.)
Mme Ben MAYER
Stephen O'MEARA (Cambridge, Mass.)
John PAZMINO (New York, N.Y.)
M. SCHEVER
Norman SPERLING (Oakland, Calif.)
Erkki VILKKI (Williams Bay, Wisc.)
Thomas WILLIAMS (Houston, Texas)
Anna Fay WILLIAMS (Houston, Texas)
Frank WOOD (Gainesville, University of Florida)

FRANCE

Dr. Agnès ACKER (Planetarium Strasbourg)
Dr. Jean-Eudes ARLOT (Bureau des Longitudes, Paris)
Bernard ARQUIER (Perpignan)
Laurette ATTIAS (Paris)
Prof. Pierre BACCHUS (Montereau)
Louis BALGAIRIES (Paris)

Joël BARBIER (Cannes)
Daniel BARDIN (Marseille)
David BARTHELEMY (Aulnat)
J.M. BECKER (Besançon)
Simon BENICHETTI (Paris)
Pierre-Michel BERGE (Les Ulis)
Liliane BERGEAL (Bureau des Longitudes, Paris)
Evelyne BLOMME (Villemomble)
Marie-Claude BONNET (Toulouse)
Christian BOURDEILLE (Gretz-Armainvilliers)
Christian BUIL (Toulouse)
P. BUTZBACK (Besançon)
Roland CARON (La Garenne)
Joseph CELO (Sceaux)
Roger CHANAL (Nurol)
Jeanine CHAPPELET (Planetarium de Nice)
Roger CHEMIN (CERGA, Caussols)
André CHENEVEZ (Garches)
Bernard CHRISTOPHE (Paris)
Alain CIROU (Paris)
Jacques CODRY (Beaumont)
Madame COLLIER (Aulnat)
Robert CONDUCHE (Argenteuil)
François COSTARD (Paris)
Philippe de la COTARDIERE (Paris)
Dr. Paul COUTEAU (Observatoire de Nice)
Mme CRAMA (Marseille)
Daniel CRUSSAIRE (Champigny sur Marne)
Catherine DAEHN-THEBAULT (Boulogne)
Germain DANIEL (Paris)
Yvon DARGERY (Paris)
Jean DAUBY (Cavalaire)
Dr. Suzanne DEBARBAT (Observatoire de Paris)
Françoise DELMAS (IAP, Paris)
Antoine DEMERLIAC (Brest)
Dr. Audoin DOLLFUS (Observatoire de Paris-Meudon)
Jacques DOLLFUS (Les Essarts-le-Roi)
Michel DUMONT (GEOS, Versailles)
Dr. Simone DUMONT (IAP, Paris)
Bernard DUPIN (Paris)
Pierre DURAND (Bourges)
François DURBACH (Pont-à-Mousson)
Yvonne ENSARGUEIX (Villepreux)
Jean-Michel FAIDIT (St. Chaptes)
Gino FARRONI (Chambray-les-Tours)
Karl FISCHER (Wissembourg)
Dr. Michèle FLOQUET (Observatoire de Paris-Meudon)
Dr. Robert FUTAULLY (OPMT, Toulouse)
Stéphane GARRO (Grenoble)
Jean-Louis GRATADOUX (Surgères)
Dr. Michèle GERBALDI (IAP, Paris)
Dr. Solange GRILLOT (Observatoire de Paris)
Abbé Joseph GRUMEL (Gap)

Yves GUIMEZANES (Perpignan)
Jean GUNTHER (Paris)
Dr. André HECK (Observatoire de Strasbourg)
Helena HEDREUL (Châtenay-Malabry)
Roger HELLOT (Strasbourg)
Dr. J.L. HEUDIER (CERGA, Grasse)
Alain HEURTEL (Besançon)
Françoise HOURMAT (Le Chesnay)
Edouard ISSENMANN (Le Chesnay)
Renée JAMMET (Rochechouart)
Dr. Pierre JONQUET (Montpellier)
Pascal JOUANNEAU (Chambray les Tours)
Marie-Claude LASTRADE (Combs la Ville)
Olivier LAS VERGNAS (Paris)
Stéphane LECOMTE (Mainvilliers)
Jean-Pierre LEGRAND (Colombes)
Michel LEGRAND (Coesmes)
Alain LE GUE (St-Brieux)
Arnaud LEMASLE (St-Etienne du Rouvray)
Roger LERISSEL (Bondy)
Hervé LE TALLEC (Toulouse)
Prof. Annie-Chantal LEVASSEUR-REGOURD (Université Paris VI)
Louis LORILLOU (Les Montils)
Michel LYONNET du MOUTIER (Colombes)
Gilbert MAHOUX (Jouy-en-Josas)
Pierre MAGNIEN (Besançon)
Louis MARQUET (Sèvres)
Jean MARTINET (Chatenay-Malabry)
Patrick MARTINEZ (Toulouse)
Dr. Marie-Josèphe MARTRES (Observatoire de Meudon)
Jean MAZEL (Béziers)
Roger MEJECAZE (Labastide-Murat)
Jean-Claude MERLIN (Le Creuzot)
Annick MERLIN
Thierry MIDAVAINE (Paris)
Joël MINOIS (Le Breuil)
Dr. Bruno MORANDO (Bureau des Longitudes, Paris)
Marie-Pierre MOREL
Marcel MOUGIN (Fronton)
Paul MULLER (Magagnosc)
Nicole NEE (Paris)
Christian NITSCHELM (IAP, Paris)
Gérard OUDENOT (Palais de la Découverte, Paris)
Dr. Jean-Paul PARISOT (Observatoire de Besançon)
Prof. Jean-Claude PECKER (Collège de France)
Bernard PELLEQUER (Observatoire d'Aniane)
Jacques PERNET (Montgeron)
Robert PERRIN (Montbeliard)
Alfred PEZZANA (Saulx-les-Chartreux)
André PHILIPPE (Wittelsheim)
Jacques PINSON (Bourges)
Dr. Dominique PROUST (Observatoire de Paris-Meudon)
Lucien REBUFFAT (Metz)

Joseph REMIS (Saint-Avold)
Dr. Jean-Claude RIBES (Observatoire de Lyon)
Pasteur André RICHARDOT (Observatoire de Sabarat)
Prof. Jean RÖSCH (OPMT, Bagnères-de-Bigorre)
Monique ROUGEOT (Besançon)
Philippe ROUSSELOT (Besançon)
Hervé ROY (Maisons-Lafitte)
Michèle ROY (Maisons-Lafitte)
Robert SAGOT (Paris)
Olivier SAINT-PE (Observatoire de Meudon)
Maurice SALAMAN (Sévrier)
Hervé SCHILZ (Nancy)
Emile SCHWEITZER (Strasbourg)
Prof. Jean SEIDENGART (Université Paris X)
Maxime SILHOL (Pertuis)
Dr. Paul SIMON (Meudon)
François SKOTARCZAK (Souchez)
Edgar SOULIE (Châtenay-Malabry)
Françoise SUAGHER (Besançon)
Richard SZCEPANIAK (Paris)
Pierre TABOURIN (Paris)
Y. TANGUY (Besançon)
Serge THEBAULT (Boulogne)
Jean-Claude THOREL (Villepreux)
William THUILLOT (Paris)
Jean-Paul TRACHIER (Observatoire de Triel)
Danielle VAPREAU (Clamart)
Jacques VIALLE (La Rochelle)
Georges VISCARDY (St-Martin-de-Peille)
Jean VODENTCHAROFF

GRECE / GREECE

Maria LASKARATOS-LADA (Athènes)

HONGRIE / HUNGARY

Janos PAPP (Urania Observatory, Budapest)

INDE / INDIA

Kunda BARVE (Nehru Planetarium, Bombay)

IRAN

Hadi HADAVI (Biruni Observatory, Shiraz)

ISRAEL

Andrei Radu SERBAN (Jérusalem)

ITALIE / ITALY

Sandro BARONI (Milano)
Marco CAVAGNA (Milano)
Pierantonio CINZANO (Thiene)
Prof. Giancarlo FAVERO (Padova)

Marco FALORNI (Firenze)
Ennio PORETTI (Merate)

JAPON / JAPAN

Dr. Ichiro HASEGAWA (Nara)
Kiyoshi KITAZAWA (Osaka)
Sei-ichi SAKUMA (Kawasaki)
Koichiro TOMITA (Tokyo)

PAYS-BAS / THE NETHERLANDS

Henk J.J. BULDER (Zoetermeer)
E.P. BUS (Groningen)
Theo JURRIENS (Groeningen)
Jan LUURS (Amsterdam)
Niels J. NELSON (Ej Hoven)
Willem NOBEL (Amsterdam)
Peter SERNE (Amsterdam)
Rob van de WEG (Borne)

POLOGNE / POLAND

Gregor DUSZANOWICZ (Cracovie)
Prof. Cecylia IWANISZEWSKA (Copernicus University, Torun)
Katarzyna TURAJ (Cracovie)
Arkadisz WAWRYSZEWICZ (Gizycko)
Dr. Krysztof ZIOLKOWSZI (Varsovie)

ROYAUME UNI / UNITED KINGDOM

Storm DUNLOP (BAA)
Andrew J. ELLIOT (Leeds)
Mart de GROOT (Armagh)
Edmund HYSOM (Cambridge)
Mrs HYSOM (Cambridge)
Prof. A.J. MEADOWS (Loughborough)
H. Robert MILLS (Salisbury)
Dr. Patrick MOORE (Selsey)
John MASON

SUEDE / SWEDEN

Lennart DAHLMARK (St. Etienne-les-Orgues, France)
Dr. B.A. LINDBLAD (Lund Observatory)

SUISSE / SWITZERLAND

Dr. Michel GRENON (Observatoire de Genève)
Kurt LOCHER (Grüt)
Eric SCHAERLIG (Radio-Suisse romande, Genève)
Andreas TARNUTZER (Lucerne)

TCHEKOSLOVAQUIE / CZEKOSLOVAKIA

Prof. Josep KLECZEK (Astronomical Institut, Ondrejov)

THAILANDE / THAILAND

Mrs. Y. VANICHAI (Nondhaburi)

U.R.S.S. / U.S.S.R.

Prof. FROLOV (Moscow)
Dr. MOZHZHERIN (Sciences Academy, Moscow)

YOUGOSLAVIE

Vladan CELEBONOVIC (Beograd)
A. TOMIC (Beograd)

GPSR Compliance

The European Union's (EU) General Product Safety Regulation (GPSR) is a set of rules that requires consumer products to be safe and our obligations to ensure this.

If you have any concerns about our products, you can contact us on ProductSafety@springernature.com

In case Publisher is established outside the EU, the EU authorized representative is:

Springer Nature Customer Service Center GmbH
Europaplatz 3
69115 Heidelberg, Germany

Batch number: 10704763

Printed by Printforce, the Netherlands